How Much Globalization Can We Bear?

How Much Globalization Can We Bear?

Rüdiger Safranski

Translated by Patrick Camiller

polity

First published in German as *Wieviel Globalisierung verträgt der Mensch?* by Rüdiger Safranski © Carl Hanser Verlag München Wien 2003.

This translation and preface first published in 2005 © Polity Press

The right of Rüdiger Safranski to be identified as Author of this Work has been asserted in accordance with the UK Copyright, Designs and Patents Act 1988.

Polity Press
65 Bridge Street
Cambridge CB2 1UR, UK

Polity Press
350 Main Street
Malden, MA 02148, USA

ISBN: 0-7456-3388-9
ISBN: 0-7456-3389-7 (pb)

A catalogue record for this book is available from the British Library.

Typeset in 10.5 on 12 pt Sabon
by Servis Filmsetting Ltd, Manchester

For further information on Polity, visit our website: www.polity.co.uk

Contents

Preface: Understanding Globalization – Between Sociology and Philosophy

Globalization is certainly one of the most widely debated topics of our time. The issue arises wherever one looks, and one wonders whether anything new remains to be said. Rüdiger Safranski's account of the issue leads to the instant, if surprising, realization that the answer to this question is most definitely 'yes'.

'Individualization' has frequently been proposed in sociological debate, as the conceptual counterpart to 'globalization'. It has often seemed that, once these processes were fully developed, all that would be left would be individual human atoms dispersed on a globe without any political, economic or cultural structures. But regardless of whether that theory is based on any good and valid observation, nobody has drawn the conclusion that suddenly emerges as evident after reading Rüdiger Safranski's exploration of the issue: globalization, if it occurs, means a radical change in the human condition. It brings human beings into direct confrontation with the world in its totality – indeed, one might say that it returns to such a confrontation, after centuries of attempts to build institutions that mediate between human beings and the world. Almost unnoticed in the broader debate, the scenario of globalization entails a return – in new and radical guise – of the time-honoured question of the ways of being-in-the world of human beings.

Globalization means that we humans, as self-relating animals, must also learn to relate to the whole. But what is this 'whole', out of which we cannot step, but in relation to which we nevertheless need to gain some distance, in order to exercise our powers of reason, our claim to make things around us intelligible? This question is the point of departure for the short, but provocative intellectual journey on which Safranski takes his readers. The adventures on the journey are plenty, and rewarding, and the author is the only guide we need. It is useful, however, to pose two questions briefly at the outset: why is it that much of the better-known literature on globalization fails to address this possible novelty in the human condition? And: how does the account that follows relate to the broader debates?

Action, reflexivity and boundaries under conditions of globalization

When it emerged almost two decades ago, the topic of globalization was a disturbing one. It questioned established wisdom both in the intellectual sphere and in the realm of political action. Associated with the diagnosis of the decline of the nation-state and the dissolution of boundaries in all walks of social and political life, it even challenged the very idea of human agency, be it individual or collective. That is to say, action seemed to presuppose not only an actor who somehow stands out from the world upon which he or she acts, but also a rather solid structure for that world, so that any intervention in it would have somewhat predictable effects. A globalized world, however, appeared at best fragmented in a disorderly way and at worst in a permanent state of flux and out of reach. In turn, the inhabitants of that world, who were previously seen as easily identifiable members of a class, nation or gender, were now seen as 'individuals' in the radical sense that they could be certain neither of their ties to other human beings nor of their own self and identity.

Such a world is, however, uninhabitable. And that insight seems to be the main reason why this early, disturbing perspective on globalization has gradually given way to a more orderly intellectual landscape. Broadly and somewhat schematically, there are three major ways of diagnosing the global constellation that started to emerge after the end of the Cold

War and, let us not forget, after colonialism (chapters 2 and 3 below address the global situation and the way in which it is usually interpreted). Most closely associated with the very meaning of the term 'globalization' are, first, the observers who hold that we are in the process of creating actual global structures for all major social practices – most importantly an effective world market for many products and a relatively homogeneous global (mass) culture. Significantly, this view is held in two versions, an affirmative and a critical one. The former is dominant among proponents of neoliberal deregulation projects; the latter points to an increasingly globalized resistance to such projects, most prominently voiced in the works of Michael Hardt and Antonio Negri.

Second and similarly consistently, other diagnoses insist on the persistence of cultural particularity in the world, often even suggesting that globalization tendencies may provoke the hardening of such cultural forms. As used to be the case with theories of nationalism and the nation-state, such reasoning is most often accompanied by the idea that cultural communities should give themselves a political form. The rise of communitarianism in political theory pre-dated the globalization debates and indeed at its outset was related solely to national communities. From the early 1990s onwards, however, this theme was integrated into a new culturalist diagnosis of the time, finding its most widely debated contribution in Samuel Huntington's idea of a 'clash of civilizations'. While this concept has rightly been criticized as intellectually and politically conservative, more innovative uses of what may be broadly understood as cultural thinking have also emerged in the context of the globalization debate, the most interesting of these probably being Johann Arnason's renewal of civilizational analysis in his recent *Civilizations in Dispute*.

Despite the richness of reasoning in both these points of view, and particularly in the latter one, what is most characteristic of the current debate is that the basic theoretical positions adopted can be criticized relatively easily on conceptual grounds. It is, after all, not very difficult either to show that numerous social practices, even many economic ones, hardly globalize at all, or to raise doubts about the idea that social life naturally occurs within relatively closed and coherent cultural containers. As a consequence, a third position has emerged and consolidated as something like a critical mainstream – for reasons I will explain, this is not an oxymoron – in the globalization debate. It might be said that this third position emerged as the globalist take on the sociological debate about reflexive modernization, most strongly associated with Anthony Giddens and Ulrich Beck (and discussed as the third variant of 'globalism' in chapter 3 below).

From this perspective, modernity was seen, according to the sociological tradition, as an institutional constellation that had triggered a particular dynamics of societal development. Deviating from the sociological tradition, however, Giddens and Beck recognized that this institutional constellation did not incarnate modernity *tout court*, but could itself undergo further transformations.[1] Significantly, some recent transformations have been seen as a reinterpretation of the modern project in the light of the preceding experiences with the institutionalization of the modern self-understanding. That is why the term 'reflexivity' has become central to this diagnosis.

Thus, the otherwise so-called decline of the nation-state was regarded as part of the general reflexive reinterpretation of modernity, even though possibly as that part that touched the very institutional pillar upon which the original modern project was founded and the boundaries by which it was protected and made viable. Rather than seeing this development as a mortal danger for the modern project, however, the theorem of reflexive modernization accepted the idea of social bonds increasingly being constructed and reconstructed through flexible networks rather than formal organization. To uphold the normative commitment of modernity to democracy hitherto incarnated in the nation-state as the organ of societal self-determination, the approach made way for the parallel revival of cosmopolitan political theory. Basically, the hope and expectation was that, if the reflexive approach was suitably understood and embraced by political actors, the newly emerging problems could potentially always be reflexively addressed and successfully dealt with.

Following its globalist approach, the theorem of reflexive modernization has become, in political terms, something like the intellectual wing of global social democracy. The position it takes is critical of both neoliberalism and 'neoculturalism', and thus of many of the powers-that-be. At the same time, however, its acceptance of the diagnosis of the dissolution of boundaries and of flexibilization (moderate though it is when compared to some other contemporary diagnoses) has also entailed an increasing vagueness in the way in which key questions of social and political philosophy can be addressed. Put very crudely, the globalist version of reflexive modernization theory marks a politico-intellectual position with which one can too easily agree from too numerous particular viewpoints because it is both broadly reasonable and at the same time insufficiently precise. That is why it has attacked many followers and has intellectually turned into a mainstream position.[2]

'Irresolvable contradictions' and the overstretched 'we': a persistent struggle for freedom, meaning and recognition

One way of describing Safranski's essay is to say that he insists that more needs to be thought and said about the challenges globalization poses to the human condition. Clearly, he, too, is strongly critical of both neoliberalism and 'neoculturalism' (see chapter 3 below). But all versions of 'globalism' – the summary term he uses for all approaches that embrace the processes of globalization – are seen by him as evading the crucial issues. Even though he does not address the theorem of reflexive modernization in any greater detail, and although there is reason to assume that he would reject it less strongly than the other contemporary diagnoses, his analysis nevertheless suggests a quite different take on the current situation. Had he used the language of the contemporary sociology of global modernity, he might have said something along these lines: it is not humanly possible to live in a widely extended world by constantly monitoring and reflexively reconsidering one's own position in it and linking up flexibly to whatever other beings and objects there are in that world. And that is why there is no gently critical perspective on globalization. Rather, we must keep asking the question that guides the essay that follows: how much globalization can we bear?

As noted at the outset, many readers who are broadly familiar with the general debate on globalization will find the interpretation Safranski offers unusual and possibly sometimes difficult to relate to. The reason is that the author found it necessary to change genre. While the globalization debate, even in its more popular journalistic forms, is inspired by sociology and political science, the particular interest and considerable originality of this essay lie in Safranski's philosophical interpretation of the question, an approach evident from the outset.

Safranski starts with a standard definition of what globalization means in terms of social processes, and then moves – as indeed do many other authors – to Immanuel Kant's cosmopolitanism (in chapter 5). But from there Safranski does not shift towards the modern state system, organized along lines of national sovereignty, and the crisis of that system today, as does virtually every other author; rather he turns to philosophical approaches, such as Idealism and Romanticism, that have their critical origins in a reaction to Kant. This is what I refer to as the philosophical rather than politico-sociological approach, and the remainder of this preface will try to explain what difference in perspective such a shift entails.

In sociological terms, the promise of globalization resides in individualization seen as the liberation from socio-institutional constraints. In philosophical terms, that same promise could be called the hope for freedom within a cosmopolitan order. Safranski accepts this idea, but from this starting point he develops a kind of dialectics of globalization – not his expression – by emphasizing that human beings are better able to open themselves towards others the more they are embedded in a world that is meaningful for them. Thus, the central philosophical issue of globalization is not the boundaries of societies and polities as such, but the individual human being's need for both freedom and meaning. Safranski derives this perspective from authors such as Rousseau, Thoreau and Hebel, who addressed the quest for a cosmopolitan order when it arose in early political modernity. And he reinterprets these themes through eyes trained by focusing on philosophers such as Nietzsche and Heidegger, that is, by observers who had already witnessed the failure of the first modern cosmopolitanism.[3]

Philosophically speaking, globalization is an 'intellectual project that seeks to conceptualize totality and to create unity'. Initially this seems to suggest that the eradication of all boundaries and the overcoming of all particularisms is to be addressed, just as in sociology and political science. The philosophical traditions Safranski bases his reasoning on, in contrast, have always insisted that human beings can be thought about only in the plural.[4] Human beings situate themselves in the world through the process of thought, and that means that they establish their own existence in relation to time, on the one hand, and to other human beings on the other. Thus, the work of consciousness consists precisely in making distinctions. Under conditions of globalization, this process takes place against the background of the reference towards totality, but this basic assumption does not eliminate the need to draw boundaries and make distinctions; rather it gives this need a specific nature.

Safranski thus allots a pivotal place in his thoughts to 'reflexivity', but he addresses this theme by other means. To reflect about the place of human beings in the world is the task of philosophy, and in particular of the tradition of philosophical anthropology with which he engages. However, any work of reflexivity – for which the term 'consciousness' would generally be used – 'results in a broken link with the world. It plunges us into time: into a past that harasses us because we cannot forget it and that remains present even when repressed; into a present that constantly escapes our grasp; and into a future that may become a disturbing scenario beset with threats.' Such a break with the world generates the most fundamental questions: it posits as central the singular human being and asks about the ways in which that human being is both always embedded in the world and assumes a reflexive distance from it.

Every such singular human being situates him- or herself reflexively in the world and will always be found at a particular position within it. The era of globalization should not be thought of as that epoch in world history in which all human beings will see themselves in the same, indistinct situation. There will always be some need for understanding one's own situation by drawing a boundary and conceptualizing 'otherness'. Rather than assuming that the condition of particularity can be overcome, and that the era of globalization is the one in which potentially it will be overcome, our attempts to grasp the contemporary human condition should aim at finding ways of thinking specificity without characterizing others as enemies to be annihilated, and ways of drawing boundaries without the assumption that any crossing of them would lead to a clash of civilizations.

In other words, striving for recognition is not a violation of the modern commitment to equality but part of the human condition.[5] And situating oneself in comparison to others should not be seen as a traditional habit of human beings that will vanish under the yoke of globalized practices. True, the current condition has led to a valuable return to cosmopolitanism in political theory. The latter contains the potential for re-constituting the political in a truly modern way, by means of 'democracy, market and publicity', as Kant hoped. But this will be the case only if we do *not* assume that democracy today demands all-inclusive, individual membership in a single global polity; or that markets today should create a homogeneous global network of competitive commercial practices; nor that the call for global publicness should level all particularity of intellectual and cultural exchange.

The need to make distinctions and to draw boundaries is often seen today as a conservative demand, and in some forms it certainly is. The claim, however, that every human being in the world could reflexively and flexibly relate to every other human being to create a homogeneous world order is not only unrealistic, an 'excessive demand'; it is also undesirable because of some features of humanness that a philosophy, rather than a sociology, of globalization can reveal. Such a philosophy of globalization would be based, unlike conservative positions, on a commitment to freedom, as Safranski's is. But freedom should not be an abstract concept, as it is in much political theory, but rather one of 'situated freedom' (as Charles Taylor would put it), for which one needs a philosophical inspiration of the kind adopted in the essay that follows.[6]

Peter Wagner
Florence

Notes

1   For my own analysis of the historical transformations of modernity, see *A Sociology of Modernity* (1994); for an argument on the need to go beyond institutional analysis, see *Theorizing Modernity* (2001).

2   The terms of these observations emerged in a discussion with Luc Boltanski, whose recent sociology of abortion, *La condition fœtale* (2004), similarly suggests that basic moral problems persist even after an accomplished individualist liberation from socio-institutional constraints.

3   See his earlier works on Heidegger (*Martin Heidegger – Between Good and Evil*, 1999) and Nietzsche (*Nietzsche: A Philosophical Biography*, 2001).

4   In the recent past, this theme has probably been most consistently addressed by Hannah Arendt.

5   On this topic see Axel Honneth's social philosophy, in particular *Struggle for Recognition* (1996), partly inspired by the same tradition as Safranski's.

6   In this respect, Safranski's approach resonates with current work, pursued by Nathalie Karagiannis and myself under the title *Varieties of World-Making: Beyond Globalization* (in preparation), which aims at moving the globalization debate away from the starting assumption about anonymous, actorless processes towards the idea of competing projects to give meaning to the world as a unity, as a globe.

1

First Nature, Second Nature

Man is a being who can relate to himself. This capacity for self-relation is precisely what is known in the philosophical tradition as 'reason', and there distinguished from 'understanding'. We encounter 'understanding' in the animal kingdom. A chimpanzee who learns from experience to fish with a stick for a banana gives proof of such intelligent behaviour. Understanding is at work where tools are produced. Animal understanding knows all about means, but ends are set for it in advance by the instincts. Reason, on the other hand, is capable of setting goals for itself. This assumes a self-relatedness that makes it possible to take a step back from ourselves and thus to survey the means–end relationship. Reason comes into play when knowledge not only accompanies but generates the act of willing – in short, when we are able to set long-term goals for which the will must first be mobilized. This requires us to step outside or beyond ourselves. The career of man as a rational being begins with this stepping out, this self-transcendence.

Man, the transcending animal, enjoys the proud distance from which he contemplates the whole; it gives him the sense of being godlike. At the same time, he notices that although he can step outside himself he cannot step outside the animal world: he is part of it. He is torn between being a god who sees the whole and an animal that belongs to the whole. But what is the whole? 'In endless space countless luminous spheres, round each of which some dozen smaller illuminated ones revolve, hot at the core and covered over with a hard cold crust; on this crust a mouldy film has produced living and knowing beings.'[1] That is a global human self-perception that can scarcely be distinguished from depression. Life discovers itself as a mouldy film on a cooled planet.

If reason is able to see the whole in such ways, the suspicion stirs in us that such reason may afflict us as a disease. Does it not demand too much? Are we not 'defective beings' precisely because we can look out on a horizon that is too wide and too distant – the horizon of the global? Is our abundance of knowledge and perspectives not our weakness?

Man is, as Nietzsche put it, the 'not yet determined animal', a semi-finished product: a being who is not yet consummated but has still to complete himself, and who has a remarkable ability to make up for natural defects with skill and intelligence. Defective being: this means that, in comparison with the rest of the animal kingdom, man's instinctual endowment is inadequate. Man is unable to rely on his instincts: he has too many options. There is too little constraint and too much freedom. When nature left him in the lurch, he had to take his own evolution in hand in order to survive. Another way of putting it would be to say that man is by nature dependent upon artificiality, and therefore upon culture and civilization. It is through culture that man, the indeterminate animal, shapes his nature, his cultural 'second nature'. In his imagination he has always been a little ahead, anticipating and rehearsing his second nature. For example, it was in religion, metaphysics and fairy tales that man made the first attempts to fly. And, since we actually have been able to fly, religion, metaphysics and fairy tales have declined in significance. In his first nature, man is a being driven by fear. Everywhere dangers lie in wait, and as man's imagination is more developed than his instincts he sees nothing but imaginary causalities in the threatening external world. To avoid being overwhelmed by his own fantasies, man had to invent cognition: he thus came to recognize, for example, that lightning was produced by meteorological conditions; it was no longer a divine judgement striking his conscience. Instead of praying, humans preferred to erect lightning conductors. The second nature that we create for ourselves is, like much else, a lightning-conductor culture.

It represents the easing of burdens, the restriction of fear and the reduction of risk. Technology enables us to create for ourselves prostheses and armour, shells and shelters.

There can be no doubt that the culture of technology and science usually stands us in good stead. But we also have problems with it – which is why, now and then, we begin to suspect that it might be better if we knew less than we do. This suspicion is as old as culture itself.

In Greek antiquity there was competition between the theoretical curiosity represented by philosophy and the art of tragedy. Plato had no use for tragedy, as its wisdom consisted in leaving certain things obscure or insoluble. Think, for example, of the drama between Antigone and Creon, as it was acted out on the Sophoclean stage: both were right, and this led to a frightful collision and a tragic outcome. A philosopher like Plato could not accept that both were right: to get to the bottom of things meant, for him, to make an unambiguous decision about the good and the bad; the true Logos knew what was right. In Plato's view, there could be tragedy only where defective knowledge blinded people to conditions that had once been apparent. Take a tragic figure such as Oedipus, of whom it cannot be said that it was good for him to discover everything about himself. Without the terrible things that he learned, he could have had a better life. The wish to know more was his undoing. A third example is Prometheus, who in Greek myth brought fire to man and thereby enabled his cultural advance. Less well known is another version of the myth reported by Euripides, in which human beings crouched idle and semi-conscious in their caves because they knew the hour of their death. They knew too much. Then Prometheus came and gave them the gift of forgetfulness: of course they still knew that they would die, but not when. By taking away that unreasonable knowledge with which they had been unable to live, Prometheus brought them relief. A new enthusiasm for work appeared among them, and Prometheus stimulated this further with the gift of fire.

Greek tragedy and myth understand that in matters of human knowledge – hence in our second nature, in culture – there can be too much of a good thing; we can demand too much of ourselves through technology and knowledge. The problem may be formulated as follows. How far can man with his second (cultural) nature depart from first nature? Can his second nature not become caught up in an opposition to first nature that is actually self-destructive?

The problem lends itself to copious illustration, but is posed with particular emphasis in the well-known case of weapons technology. It cannot be denied that our basic emotional equipment goes back a very long time; it belongs to first nature, as it were. This is also true of our aggressive potential. But, instead of clubs, which can only reach so far, we now have modern weapons technology. One touch on a button and hundreds of thousands die. We can scarcely imagine the effects of what we do; we are more capable of producing than of imagining. This is all well known and the object of much reflection, although obviously that does not remove the problem. And today this tension especially affects the problem of globalization.

In the last few decades globalization has gathered tremendous momentum, but its prehistory stretches further back. Here are a couple of examples:

> Modern industry has established the world market . . . Constant revolutionizing of production . . . distinguishes the [present] epoch. . . . All fixed, fast-frozen relations, with their train of ancient and venerable prejudices and opinions, are swept away. . . . All that is solid melts into air, all that is holy is profaned.

This we read in a text from the nineteenth century. At approximately the same time another author wrote:

> Now any little country town and its surrounding area, with what it has, what it is and what it knows, is able to seal itself off. Soon that will no longer be the case; it will be wrenched into the general intercourse. Then, to be adequate for its contacts on every side, the lowliest will have to possess much greater knowledge and capacity than it does today. The countries which . . . acquire this knowledge first will leap ahead in wealth and power and splendour, and even be capable of casting doubt on the others.

Both texts come from the middle of the nineteenth century: the first from the *Communist Manifesto* by Marx and Engels; the second from the novel *Der Nachsommer* by Adalbert Stifter.[2] One is revolutionary, the other conservative. In each perspective, however, modern globalization has already begun and is experienced in contemporary consciousness as the beginning of a future which, as we know, has never stopped beginning.

The man-made network of second nature, which covers our blue planet like a mould, is a global network. The first globe was produced in the sixteenth century, in Nuremberg. From that time there was a material pendant to global consciousness, if only in the form of a model; we were able to take this model world in our hands. Nearly 500 years later, cosmonauts were able for the first time really to look at our world as a globe, and since then we have grown used to that image as we do to all others. The moon landing in 1969, with its view from space over this blue planet of ours, was probably the moment at which modern global consciousness was born, the beginning of the fall from euphoria into panic. For, as world society and world history were wrenched as never before into globalization, the apprehension grew that there might be too much globalization or too much false globalization, and even that we might be on the wrong track altogether. The doubt and the unease eventually led to an anthropological question: how much globalization can we bear?

2
Globalization

First a few facts.

We live in an age of globalization – no doubt about that. Since the atom bomb there has been a globally shared threat. Missiles can reach any point on earth. The nuclear potential makes collective human suicide and global devastation a real possibility; life on earth is on the line. Wars are no longer limited to particular regions, nor conducted only by states. Non-state violence, or terrorism with bases of support in a changing list of states as well as links to organized crime, seeks to mount global operations and to procure weapons of mass destruction. We have known this since 11 September 2001, but even before we had to fear it. The diversion of nuclear energy from civilian to terrorist purposes – for example, through an attack on a power station – is possible at any time. And other dangerous technologies now in civilian use, such as biological or genetic engineering, could be given a terrorist application – with global effects. These few words may be enough to remind us that modern globalization began with the globalization of fear and terror.

This also holds in relation to ecology. Economic and industrial overexploitation, on land, air and water, is condensing into a truly frightening backdrop. Globalization in this sense means the plundering of our planet.

We could continue the list of horrors associated with, or directly traceable to, globalization. Diseases follow in its wake, either in reality or in people's fantasy. Aids has been transforming the world into a global community of infection. And overpopulation is another terrifying aspect of globalization.

Then there are the more narrowly economic and technological processes of globalization that increase the density of economic, cultural, touristic, scientific, technological and communicative networking. According to an OECD definition, economic globalization is the process whereby markets and production in various countries become increasingly interdependent as a result of cross-border trade in goods, services and labour and the movement of capital and technology. Globalization signals the triumph of a capitalism which, since the collapse of the Eastern bloc, has been the one dominant economic model. Despite the persistence of political and religious differences, the forms of economy and technology have become more unified, albeit at different levels of development. There are counter-tendencies, but even those are dependent upon capital and Western technology. Deregulation of financial markets spells ruin for entire economies. Corporations with global operations disempower locally legitimated politics. Capital flows like a river across national frontiers, and causes flooding and proliferation, desiccation and drought, not only in the metaphorical sense. The total effect is like that of a worldwide natural disaster, man-made though unplanned. Yet the whole thing unfolds with the help of precise technologies and calculated strategies of profit maximization, rational in the particular but irrational overall. Nor does public opinion escape attention. Global information technology makes it possible for us to know anywhere in the world what is happening anywhere in the world.

This high degree of self-reference and visibility is part of modern globalization. Everything has always been connected with everything else. But the processes used to operate behind the backs of those involved in them, whereas now a perception of networking that wavers between euphoria and hysteria has become part and parcel of globalization. Thus modern globalization means, first, networking as such and, second, networking that people know about – or rather that they know of, since all they know for sure is how much everything is bound up with everything else. Modern globalization is self-referential and – as far as communications technology is concerned – takes place in real time.

Now, globalization is not only the quintessence of the bad news travelling around the world; there are also nice enough communities engaged in global cooperation. It cannot be denied that the spread of modern science, medicine and technology has effects that make life easier and safer. At an institutional level, the United Nations and a dense network of complementary (or rival) intergovernmental and supranational organizations and agreements exist to keep war and violence in check. A world public opinion has established itself, so that tyrannical regimes are made to feel under scrutiny and legitimation pressure. Human rights abuses, though not systematically punished, do not pass unnoticed but give rise to worldwide protests. An international criminal court is being set up. And critics of globalization form a growing movement that makes expert use of the global technologies of information and mobilization.

3
Globalism

'Globalization' is not a single process but has several different facets, and for this reason it is better to speak of 'globalizations' in the plural. Nor is there any shortage of alternative (counter-) globalizations: that is, attempts to control and reshape the dynamic of capital and technology. If critics of globalization – governments, organizations or individuals – link up their economic and informational networks, their aim is to gain new scope for action and to develop alternative forms of a global ethos. This mostly happens at a practical level, but the scale of the global tasks also favours the emergence of broad ideological theories of the global, with either a critical or an affirmative intent. Thus, we have to deal not only with factual globalization but also with 'globalism' as an idea or ideology.

Globalism qua ideology generates the image of a world society more unified than it is in reality. Often it suppresses the fact that, while some regions are becoming more homogeneous, others are experiencing a dramatic de-linking from events in the rest of the world. Certain societies and regions communicate with one another, but others become 'blank areas' and regress to earlier stages of development. In a world that communicates in real time, lack of simultaneity is a growing phenomenon. New time zones are taking shape – not in the sense of the clock but in the sense of different epochs. In Africa, for example, countries are dissolving into tribalism and gang warfare; feudalization, robber barons and pirates are making a comeback; unimaginable poverty and a brutish struggle for survival are cancelling the rules of social existence. The very minimum of civilization is disappearing.

The dramatic lack of simultaneity causes various reflexes among the public in the West. Everything bad that it thinks to be impossible over here is considered possible in the East – for instance, a nuclear war between Pakistan and India. Those who try to avoid alarmist panic stick to the cosy liberal but now also cynical axiom: other countries, other customs. China's strict policy of one child per family may be praised for its dampening effect on population growth, and at the same time despised because it contradicts our democratic standards.

Thus, new differences emerge on the ground of globalization, but not many suitable forms of behaviour to handle them. Globalism qua ideology does not want to see the growing lack of simultaneity and the differences in development, or else it treats them only as transitional phenomena. This means that it is not sufficiently realistic. At bottom, globalism is less a description of reality than a demand: it makes a global 'ought' out of a global 'is'. Globalism is globalization become normative. If it remains undogmatic, flexible and rich in insights, then it is a question of ideas – but otherwise it is ideology. In one way or another, globalism reacts upon the real movement of things, by disguising, forcing, crippling or legitimating. As an ideology, it is the intellectual side of the global trap.

Three variants of normative globalism may be distinguished. First there is neoliberalism, the most effective variant. Because it is so powerful, it is the most likely to be denounced by the critical public. Neoliberalism invokes globalization as an argument for ending the social obligations of capital, and counts on competition among governments for jobs to eliminate so-called investment barriers (by which is meant ecological and fiscal regulation, social welfare policies and legislation favourable to trade unions). Neoliberal globalism is an ideological legitimation for the unrestricted movement of capital in search of the most favourable profit conditions. In order to impose the primacy of economics over the state and the local culture, it brandishes a threatening scenario in which the flow of capital is switched off. Neoliberalism is as economistic as vulgar Marxism used to be; it is therefore in a sense the resurrection of Marxism as a management ideology. In effect it aims at a world order already described in the *Communist Manifesto*: 'it has put an end to all feudal, patriarchal, idyllic relations. It has . . . left remaining no other nexus between man and man than naked self-interest, than callous "cash payment". . . . It has resolved personal worth into exchange value.'[1]

Ideologies of the unfettered market are readily understandable in terms of the interplay between 'is' and 'ought'. It is explained that economic being determines consciousness, but also demanded that it should determine consciousness as nicely as possible. The interplay between factual and normative has certain advantages: one can deflect a normative critique of the market by falling back on the power of the factual, or counter an empirical critique of market realities by appealing to the supposedly unfulfilled idea of the pure market. This makes it possible to occupy the fields of 'is' and 'ought' at one and the same time, with references to Adam Smith as the great theorist of the market. He wrote: 'He is not a citizen who is not disposed to respect the laws and to obey the civil magistrate; and he is certainly not a good citizen who does not wish to promote, by every means in his power, the welfare of the whole society of his fellow-citizens.'[2] The market, then, can produce the common good only if it is based upon the morality of the common good; it cannot by itself create the spiritual-moral prerequisites for this to happen. Adam Smith, himself a moral philosopher and not just a market theorist, understood something that many of his ideological progeny have forgotten for reasons of self-interest. Nobel prizewinner Joseph Stiglitz, an economic liberal and former chief economist at the World Bank, has recently called to account the neoliberal ideology that prevails in the upper echelons of that institution. He describes how the 'mantra' of open markets, privatization and social spending cuts is mechanically thrown at a complex reality, and how this can ruin whole economies, as it did in Argentina or Russia.[3] Think ideologically, act globally: that is an especially disastrous variant of 'globalism'.

A second aspect of ideological globalism is anti-nationalism. The idea that the future will be global is meant to help us overcome the traumas resulting from the destructive history of nationalism in Europe. In this context, globalization carries the sense of 'never again nationalism!' Declarations of belief in globalism are especially fervent in Germany. This is where politically well-meaning people chose first Europe, then the whole world as their refuge from a disagreeable nationalist past. Such globalism is also (like neoliberalism) normative, though with a cosmopolitan intent. Linked transnationally from above and from below, supported by technologies of communication and transportation, we are supposed to free ourselves from national incubators and archaic states of excitement. Doubtless that is thoroughly desirable – one need only think of the various unhealthy excrescences of nationalism. Yet anti-nationalist globalism does nothing to change the basic anthropological point that mobility and openness to the world need to be balanced by firm local attachments. We can communicate and travel globally, but we cannot take up global residence. We can live only here or there, not everywhere. The German language has a fine word for this emphatic sense of place: *Heimat*. But it is precisely in Germany that this expression sets alarm bells ringing; people suspect that what lies behind it is backwardness, a longing for roots, or even political revanchism. Of course, there is nothing strange in the fact that talk of *Heimat* lost its innocence for a time among a people like the Germans, who, with their battle-cry 'a nation without space', devastated the homelands of other peoples on a grand scale, or drove them from their homeland and then had to suffer expulsion and destruction themselves. We may hope that it was only temporary, however, as we now again need a positive valuation of *Heimat*, for anthropological reasons alone. The more that local attachments are emotionally satisfied, the greater is the capacity and willingness to be open to the world. The hysteria of athletically mobile 'global players' should not be confused with what is known in German as *Weltläufigkeit*, knowledge of the world based on wide and sensitive travelling. That requires a willingness to become entangled in a foreign dimension. Someone can be *weltläufig* only if they have been changed by rich experience of the world, not if they merely conduct global business or travel the world as a tourist.

Third and last there is the globalism with which modern thinking about globality really began. We look with sympathy and alarm, as if from outer space, at the poor earth that we are in the course of destroying and that we must do something to save. We discover the earth as a global biotope, as the house of our being, which technological hubris threatens with destruction unless we wake up to our common responsibility. This globalism knows how to adopt a solemn tone – the great 'we' of humanity here celebrates its resurrection. At the same time, it conjures up the horsemen of the Apocalypse to urge a turning back; prophets of doom condemn consumerist culture and keep penitential robes at the ready. In its more sober variants, this globalism involves a cool appraisal of the effects of technology and of the main risks facing the world. Its impact is beneficial because public awareness of such risks politicizes decisions about research, technology and capital investment – spheres that have until now remained largely outside the democratic process.

Nevertheless, the politicization pressure stemming from global environmental problems – one only has to think of possible climatic disasters – does not affect a single global subject who might be called to reason or held retrospectively to account. It is true that people talk of global society as global communication. But this communicating global society does not constitute humanity as an active subject – in the way in which some philosophies of history used to dream. Only individual states or alliances of states have power. 'Humanity' has no power: it is an incantation in the arena of the real powers, where global asymmetries of power, productivity and wealth generate a new type of sovereignty trap; for it turns out that he is sovereign who can pass on to others the costs resulting from his own action. In this respect, the United States is more sovereign than other states when it sabotages international agreements on the environment. If energy, water or atmospheric resources start to run out, power is always what determines the distribution of life chances. The consequences of scarcity are borne first of all by the weak – until the strong too are affected by it. It is anyway a delusion to think that global problems of an apocalyptic magnitude might lead to global solidarity. Here too it is the last who will carry the burden. This will remain the case so long as others hope they will be among the next to last.

4

Making Enemies

When we realize that, despite the shared fate of world society, 'humanity' does not exist as the urgently needed subject of action, we at least dream of such a subject and eventually demand that our dreams become reality. This is what happened especially in the eighteenth and nineteenth centuries, at a high theoretical and imaginative level. With the Enlightenment and the revolutionary foundation of democratic polities, first in America and then in France, the idea spread that the free spirit must be able to have its way with history. History was capable of being made; it had to stop being mere destiny. From Hegel to Marx, thought circled around the possibility of overcoming 'naturalness', that blind, irrational contingency which constituted first nature but must no longer determine man-made second nature. 'Humanity' was supposed to emancipate itself and gain control over its own history. The socialist project was directed at the achievement of this goal through class struggle.

But there was also a more thoughtful variant of this idea. After the collapse of 'actually existing socialism', this milder vision is appearing once again. For we dream that the world is becoming an oecumene, a single secular church council for the purposes of reflection, wise rule and a global change of course. This we might call ecumenical globalism. None other than Friedrich Nietzsche dreamed of it and thought that it would become reality in the twentieth century:

Since man no longer believes that a God is guiding the destinies of the world as a whole . . . , men must set themselves ecumenical goals, embracing the whole earth. . . . If mankind is to keep from destroying itself by such a conscious overall government, we must discover first a knowledge of the conditions of culture, a knowledge surpassing all previous knowledge, as a scientfic standard for ecumenical goals.[1]

That Nietzsche of all people should have dreamed of a global oecumene is especially surprising, for he never tired of emphasizing that there had so far been a 'humanity' only in the form of a drama involving the making of enemies. He was certainly right about that.

I do not wish here to follow the trail of blood through human history; it is well known and yet not known well enough. As usual, it has to be explained why philosophers ascribe to man a reason that is, so to speak, directed a priori at consensus. It will be necessary to reveal, at the heart of universalist concepts, the energies that have gone into exclusion, into the making of enemies. Here I can do no more than intimate that enemy-making and boundary-drawing have always been a necessary part of intellectual projects that seek to conceptualize totality and to create unity (and therefore of earlier forms of globalism). Almost every attempt at totalization sinks back into the history of enemy-making, into a struggle either against rival totalizations or against the stubborn refusal to fall into line with an ostensibly good whole. Enemy-making energies also lie hidden within theories that seek to show how peace might be brought to the whole.

For a long time God was the name that epitomized the whole and the global. But before God became universal he was a god for particular tribes and societies – a separate god, not a species god. He was therefore constantly drawn into conflicts among hordes, tribes and nations. Not a god of the whole, but a god for the rough stuff on the margins of each community.

Let us take the God of the Old Testament. He forms an alliance with his people – against the rest of the world. After the Exodus begins the conquest of the promised land, with God's encouragement: 'I will send my terror in front of you, and will throw into confusion all the people against whom you shall come, and I will make all your enemies turn their backs to you' (Exodus 23: 23). In setting himself up as the defender of unity, this god was actually the problem for which he claimed to be the solution. For he defined himself as the enemy of the gods of other tribes and peoples.

What happens when men ignore their jealous gods and strive for boundless unity of the human race is related in the story of the Tower of Babel: 'Come, let us build ourselves a city, and a tower with its top in the heavens, and let us make a name for ourselves; otherwise we shall be scattered abroad upon the face of the whole earth' (Genesis 11: 4). This bold venture, which he saw as wicked presumption, again brought God on to the scene. A higher unity could evidently exist only in God. If humans wished to attain it, they would have to deal with a god who sowed confusion in human speech, and that would mean even more of the dispersion and mutual hostility that the Tower of Babel was meant to overcome.

If the power of God and the gods over men diminishes, then it is the human species that is stuck with the task of achieving unity. This is why, after proclaiming the death of God, Nietzsche had to start thinking about the possibility of a human oecumene. But the force of the Tower of Babel as a portent of human animosities may be seen in Nietzsche's own development; he soon lost sight of the ecumenical task and turned to a kind of spiritual aristocratism whose new enemy-images involved a targeting of 'the degenerate'.

What is the source of the hostilities that divide the human species even when potentially catastrophic problems have made unity a matter of such urgency?

In former times, the making of enemies seemed so natural that we might almost say it had a priori status. This may be seen not only in the history of religions but also in the history of philosophy, which has always kept a lookout for possibilities of peace in human affairs, though only a peace within boundaries secured against a hostile, barbarian environment. The a priori status of enemy-making in early philosophy is apparent in Aristotle's image of an army fleeing an enemy, where the aim of philosophy is to make the army turn and check once more whether flight is the right course of action – that is, make certain of its own foundations (also known as principles). For it may be that the flight is groundless, or else a flight from the groundlessness of an abyss; it may be, in other words, that the enemy is the form of one's own groundless abyss. However the history of the philosophical viewpoint may end, its prerequisite is the elemental making of enemies. That is where everything began. But why is this making of enemies so elemental?

Let us try an explanation which does not, as such explanations frequently do, begin with the animal realm (that is, with man as a herd animal whose group behaviour is instinctual) but with the dimension of consciousness that sets the human animal above and apart from other animals. Consciousness results in a broken link with the world. It plunges us into time: into a past that harasses us because we cannot forget it and that remains present even when repressed; into a present that constantly escapes our grasp; and into a future that may become a disturbing scenario beset with threats. Everything would be simpler if consciousness were only conscious being. But it breaks loose and becomes open to a horizon of new possibilities. Consciousness is able to transcend the given reality, and hence to discover either a dizzying nothingness or a god in which everything comes to rest. In fact, the suspicion will not go away that this nothingness and this god may be one and the same. An animal of consciousness that can say 'no' and experience nothingness is also capable of choosing annihilation, perhaps for reasons to do with *horror vacui*. When the philosophical tradition refers to this precariousness of the human condition, it speaks of 'defective being'. Man qua conscious being has lost his security in the here and now.

This is what the biblical myth of the Fall is telling us. The expulsion from Paradise is the point at which worry is born. It is no longer enough to make ourselves secure in the present; we must also look to a future that carries so many dangers – from our own inconsistency, from other people and from an overpowering nature. The temporal horizon is enticing but also threatening. Only because there is a future threat does the desire for power arise. For it is power that is supposed to make the future safer.

Since the main point in the end is to keep the boundless future in check, and since there are others who seek to acquire power for the same reasons, power must also become unbounded. It must become dynamic. Power needs more power. It can survive only through accumulation and expansion. Power feeds on control, the act of overpowering breeds violence. Power becomes ceaseless longing. Power begins with self-preservation, and therefore with the desire to preserve over time a self that is afflicted with awareness of its own temporality. If everyone tried on their own initiative to preserve their self over time, this would inevitably lead to a war of all against all. The cultural process therefore siphons off, as it were, the individual force of self-assertion, binding and socializing it in laws, rituals and institutions. Eventually this turns into the state monopoly of violence. It makes the many individuals equal, in respect of their renunciation of violence and the associated loss of sovereignty. Yet there remains a dramatic contradiction that makes it impossible for history to come to rest in the power crystal of the sovereign state.

Contradiction is another elemental need that has its basis in the fact that man is a conscious being. Having consciousness of himself as well as an ability to compare himself with others, he must be different from others in order to be a self. Descartes's formula 'I think therefore I am' does not at all satisfy the desire for self-certainty. I may be certain in thought *that* I exist, but I learn *what* I am only in distinction to others. Even then, I cannot rest content with the discovery of difference; it must be a difference in my favour. The intense feeling of self grows out of an awareness that we differentiate ourselves by 'showing off'. It comes down to a matter of standing. Man only really enjoys what sets him apart from others.

This desire for difference gives society its dynamic, but also puts it in danger. Plato called *thymos* the passionate craving for difference, an inner force expressed in the brave action through which an individual seeks to stand out as superior. This inner force may also inspire a will for power – a will to overpower. It kindles the ambition to be master. The thymotic passion carries man beyond mere self-preservation to the pursuit of self-enhancement. Now it is a question of recognition and increased status. For the enhancement of life, the thymotic passion is even prepared to risk ordinary life.

For Hegel, this life-and-death struggle for recognition is the driving force of history; it is a dynamic not of wild human nature but of the spirit. The life of the human spirit is the triumph over death and fear of death in struggles for self-enhancement. We wish to be recognized as superior by virtue of a greater preparedness for risk. For it is true that whoever risks his or her life will win it as a higher life. The historical reality in which the concern for such recognition operates is the bloody struggle for sometimes trifling goals: people risk their lives to adjust the line of a frontier, to defend a flag, to obtain satisfaction for an insult. The kind of vulgar materialism now so influential in consumer societies discovers purely economic motives everywhere and is blind to the enemy-making energies that derive from the thymotic passion. Poverty suffers; it does not fight. Only wounded honour or the desire for glory and recognition fight. This too shows that man is an animal of consciousness, who lives not on bread alone but also on honour and dignity. That is why he fights, and that is why he has and seeks enemies. Not only peace but also war is made for the sake of 'human dignity'; so it is that competition, struggle, war and the making of enemies never cease to operate in the arena of history.

The question that has always been posed in history, whether explicitly or implicitly in people's lives, is how thymotic energy can be organized as a productive force. When such energy is bound, it gives a society its dynamic. But when it is unbound – and it always strives in that direction – it dissolves society into the anarchy of violence. Economic, political or sporting competition is certainly a well-tried way of harnessing and giving expression to the thymotic passion. If this harnessing does not work well enough internally, then – as a glance at history will tell us – there is always the less pleasant solution of transferring the struggle over difference and recognition out to the borders of society, where the world of foreigners (now declared to be enemies) begins.

The struggle for recognition aims to establish difference in one's own favour. But we should not forget differences resulting from the accident of birth, genetic make-up, family, social milieu or historical period, which used to be thought of as dictated by Fate or as God-given. In fact, from the point of view of a sense of justice, these unequal chances in life and development can only be experienced as hurtful. Thus, in addition to the differences that people establish among themselves, there are differences that count as accidental, natural or God-given; these are not made by men and women but are made together with them. It is difficult for an ugly or sick person to avoid thinking of his or her fate in terms of an unfair disadvantage. The outcome may then be envy and outrage – a further element in the making of enemies. Nor does the outrage stop at the image of a just God: for how can God be just if he permits such inequalities? This is the kind of doubt that can bring about the collapse of religion and metaphysics. The biblical myth of Cain and Abel absorbs this doubt and relates an archetypal tale of outrage and enemy-making.

The story of Cain begins with the distinction that God makes between the brothers. The scandal of the original inequality of treatment triggers the catastrophe. Cain, the farmer, sacrificed to his God from the fruits of the field; Abel, the shepherd, offered up animal sacrifices. But why did God look kindly only upon Abel's animal offering? Why did he reject Cain's sacrifice? Cain could see no reason for it. It must have seemed to him that God was refusing to treat him equally. 'So Cain was very angry, and his countenance fell.' And so God reproached him: 'If you do well, will you not be accepted?' But how can one remain 'accepted' in the face of such rejection? There can be no doubt that, in abolishing the symmetry of sacrifice and divine favour, God destroyed the harmony between the brothers. God's inexplicable behaviour caused Cain to fight against the way in which he had been differentiated. First he was the victim of difference, and only then did he commit the crime. After Cain had been so hurtfully differentiated from his brother, he established an irrevocable difference by killing him. He acted out of wounded honour, envy and bitterness against a God who had flagrantly violated the principle of justice. It appeared as if God recognized his own share of responsibility for the fratricide, for although he banished Cain – 'you will be a fugitive and a wanderer on the earth' – he also took him under his protection. 'Whoever kills Cain will suffer a sevenfold vengeance' (Genesis 4). That is how the angry and envious Cain could become the second ancestor of the human race.

The realization dawned on Adam that he would be a creature of troubles. Cain, his son, had recognition denied him, and he responded with murder. Concern for self-preservation, struggle for recognition, anger at unequal treatment: these elements combined to produce an explosive effect, making human beings at once creative and more dangerous. The old stories tell of the birth of freedom and the awakening of consciousness and a sense of time, through which the world first became an object of concern. They tell of the dramatic complications arising from the fact that people are different and react to their difference with satisfaction, outrage or envy; that they compare themselves with one another and then seek to attain, or to defend, greater balance or differentiation in their own favour. In the end it always comes down to struggle and war, which can be contained but not eliminated.

Always, however, a longing for unity accompanies the (sometimes deadly) struggle over difference. This longing is bound up with the vision of a God who towers above all difference and opposition, and with the vision of a humanity that does not know, or no longer needs to create, any difference. These are the two great ideas of universality: one God and one humanity. But, in reality, the idols are many and humanity is torn apart by different families, tribes, peoples, nations and kingdoms – and this remains the case in the age of globalization. The political world is still not a universe but a 'multiverse'.

Already Plato had to reckon with such a political 'multiverse'; we also find in him the idea – no more than the idea, of course – of the unity of the human race. The creation myth in the *Timaeus* relates how the idea of humanity was gradually watered down and polluted. The Demiurge created a basically good substance, but then so many chance accidents occurred that nothing could turn out right. The outcome is a world in which there are enemies, barbarians and wicked people. We have to learn to live with this reality. The problems that Plato had with this are reflected in the fact that his philosophy generally treats multiplicity and historical becoming as symptoms of defective being. Multiplicity and becoming inevitably mean struggle and war. Hell's denizens specialize in the culture of strife. Will humanity as a whole ever extricate itself from this Platonic hell? Hardly. Therefore the phantom pain of an impossible unity lingers on. Plato's philosophy develops out of these premises, both involving itself in the struggles of his time and rising above them in grandiose visions.

Human beings create mutual boundaries in embodied reality: they run up against one another in space and differentiate themselves. How can the Other not be mistrusted if it is already so difficult to trust oneself? How can agreement be reached with others if agreement with oneself can scarcely be achieved? How should one endure strangers if one can hardly endure oneself?

Plato's philosophy shows the art of reaching agreement with oneself. It thereby helps to stem the flow of enemy-making energies. It cannot obtain this result from the world, because boundaries are part of embodied reality. Even Hegel, a specialist in thinking beyond boundaries, maintained that in actuality they were necessary to life:

> A thing is what it is, only in and by reason of its limit. We cannot therefore regard the limit as only external to being which is then and there [*Dasein*]. . . . Man, if he wishes to be actual, must be-there-and-then [*dasein*], and to this end he must set a limit to himself. People who are too fastidious towards the finite never reach actuality, but linger lost in abstraction, and their light dies away.[2]

Just as, in Plato, the knowledge of limits brought insight into the inevitability of war, so in Hegel the necessity of boundaries for life leads to the actual dialectic of mutually contending oppositions. These oppositions sound harmless enough in the abstract form of thesis and antithesis, but in reality this conceals the struggle for life and death. In the war of the dialectic, the synthesis is most often the thinly veiled triumph of one side, who gains mastery through the Other's act of appeasement. But the victor does not remain what he was before: he takes something over from the loser, changes it and, in so doing, is changed himself. World history is a history of irresolvable contradictions, which will be fought out as long as there are winners and losers. The whole can be conceived but not lived. It can be lived only in the fighting out of contradictions, in the history of enemy-making. The whole may be the true and the good, but it takes shape only through the struggle of particularity.

Since Plato's time, then, an ever-shifting mode of thought has lost itself in the dream of the whole, without ever being able to escape the fact that people differentiate themselves from one another and actively fight out their differences to the point of life-and-death struggle. This is why, in Plato, the elevation of the spirit begins with the need to secure boundaries against the barbarian; and this is why, in Hegel, the spirit becomes real only through struggle and war. The dialectic of actuality is a bloody matter. People always exist within wrenching, antagonistic oppositions; it is part of the contingency of existence that they are born into one or another side of the contradiction. They do not choose their place but can only take it as given. The question is therefore not posed as to whether it is the 'good' side. In fact, the reverse logic applies: one side is good because I belong to it and because it is where 'our people' are. We and the others – that is an obvious differentiation, after which all that is necessary is to clarify the boundaries of the We. Can the We be limitless? Can it really encompass the whole of humanity?

There was an age which knew how to speak enchantingly and with boundless enthusiasm of the whole and humanity. It was the *Sturm und Drang* period of German Idealism, with Friedrich Schiller as its chief brain. He sang to 'Joy', whose 'magic reunites that which custom's sword has parted. Be embraced, all ye millions! With a kiss for all the world!'[3] One thing is certain: such enthusiasm is most unlikely to have developed amid the milling crowd, in the constant press of communication with his fellow humans. Only in well-measured doses can people be stirred by enthusiasm. It may be that the Idealist enthusiasm for humanity could flourish only in small towns such as Tübingen or Jena, and only in an age when there was no such thing as global communication in real time, for such enthusiasm can develop only where there are sufficient spaces for the imagination. If the spaces are too narrow and the human entanglement too dense, 'humanity' becomes no longer a dream or a distant beloved but a zoological concept referring to an overcrowded human park or a human wilderness. On closer inspection, however, what really mattered for German Idealism was not the 'humanity' of the many but the 'humanity' of the individual – that which each person discovers, respects and perhaps even loves within himself or herself.

The memory of myths and the labour of thought regularly come up against the elemental conditions of enemy-making.

We reach back into early times to grasp the moment of unity, but, like the horizon as we approach it, it keeps receding and we never find ourselves outside the history of enemy-making. Plato was content to demonstrate the conditions of possibility for a pacified zone within a hostile world; he already saw enemy-making as an a priori of history. Peace is the impossibility that we cannot give up in our thoughts and our desires. Yet any peace project that is meant to become reality turns into a bone of contention among the contending parties. It is not so surprising, therefore, that Jesus said he had come to bring not peace but 'a sword' (Matthew 10: 34) or 'fire' and 'division' (Luke 12). Swords would be turned into ploughshares only once they had done their work. The only peace that Augustine could imagine for his *civitas civilis* was an armed peace capable of defending itself – among Christian peoples, but all the more in relations with heathens, with whom peace could only be an exception. Usually it was necessary to wage war against them, a 'just war' (the expression is already found in Augustine). There could be true peace only in God, but so long as the *civitas dei* projected into the *civitas civilis*, a contested boundary would inevitably separate the two 'kingdoms' and also cut through the inner world of the individual. War was the father of all things, and that evidently included things of the spirit.

Only if we stand right back and contemplate our planet from outer space does the big moment come for the conciliatory, almost compassionate thought: 'Why do we get so worked up? Can't we put up with one another on this poor little earth on which we all live together?' Seen in this way, the earth is something to which we return with a gesture of detached and magnanimous solidarity. We find a classical formulation of this in canto 22 of Dante's *Paradiso*:

> I turned my eyes back through every one
> Of the seven spheres and saw the globe which looked
> Such a miserable thing that I smiled.[4]

Today such gazes are familiar to us from space travel, and they have led quite a few astronauts to become devout ecologists. But when ground contact immerses us again in the jungle of the social, the peacefulness up there is soon over. Besides, the militarization of space is continuing, and up there too we will find traces of global enemy-making.

Let us be clear. Looking at global problems from a great distance, we speak of the tasks and failings of 'humanity', and so on. Then we say 'we' and mean the whole of humanity. This overstretched 'we' counts as the collective subject to which an action can be ascribed. But it is only in views from above that there is a singular 'humanity' capable of action; in reality there are only human beings in the plural. It flies in the face of all historical experience to assume that, out of the human crowd, 'humanity' can take shape as a subject of action. Behind any power that sets itself up as humanity in action will always lie a particular power that uses the manoeuvre to gain an advantage for itself in the competition with other powers. Carl Schmitt's formulation 'Anyone who says "humanity" is lying' is not so wide of the mark, provided that bad intentions are not assumed in every such lie. For it is characteristic of 'humanitarian' politics that it is usually well intentioned.

5
World Peace?

It was Kant who, in the eighteenth century, outlined a conception of world peace on an underlying assumption of multiplicity. As no subsequent vision of world peace can be compared to Kant's in conceptual richness or realism, we shall take it here as representative of the whole genre. Kant too began with cosmic speculations: he asked why the inhabitants of this lost planet circling in dark space had found it so difficult to get on with one another, and whether there were not perhaps ways of changing this; any idea that human beings might actually destroy their planet could not yet have occurred to him. The possibility of global peace is the theme of his celebrated treatise that appeared in 1795 under the title *Zum ewigen Frieden*.[1]

First of all, the earth for Kant is a single global trap, because the people living on it as 'a globe . . . cannot disperse over an infinite area, but must necessarily tolerate one another's company.'[2] But how is this supposed to happen, given that there have always been wars as far back as our historical memory stretches? A degree of peace has been achieved within individual countries, but between states and between nations the principle of force and war still prevails as the *ultima ratio*, or the *ultima irratio*. It is precisely in 'the relations which obtain between the various nations' that 'the depravity of human nature is displayed without disguise.'[3] Can reason change anything in that? Kant starts from people as

they are, not as they might be. A place must be kept for
egoism, self-interest and assertiveness in any construction of a
better state of affairs. Kant points to the state's monopoly of
force within countries, which shows that it is in the individual's own interest (properly understood) to submit to rules
that restrict his sovereignty but provide him with security
and protection. Are such rules also conceivable between
nations? In other words, is it possible to overcome the 'state
of nature' not only within countries but also between them?
One possible scenario is the violent subjugation of nations
and states by a single world power (for example, the Roman
Empire in the first few centuries AD). But that is not satisfac-
tory, precisely because it involves pacification through subjugation. Another possibility is a complex policy aimed at a
balance of powers but, instead of creating genuine peace,
that only postpones the state of war. If the purpose is to end
war between countries as much as the anarchy of violence
within them, then the numerous existing states must merge
into a single world state. When a central power, however legitimated, defines and supervises a single global internal policy
and enjoys a global monopoly of force, it should be possible
to restrict and eventually eliminate international wars, just as
regional state authorities have done in the case of civil wars.
A global internal policy would mean that all wars would
count as civil wars.

It is a sign of Kant's realism that he considers such a world
state to be neither possible nor desirable. As regards its impossibility, he argues on the basis of experience: states with a
'lawful constitution' will hold on to this legitimation of
their sovereignty, and will not be prepared to relinquish the
monopoly of force which derives from and safeguards that
sovereignty. But a central world state would anyway be undesirable, because it would pose a threat to the productive diversity of nations and to their distinctive 'languages and religions'.
Such diversity and difference are part of the human domain.
Kant concedes that their preservation 'may certainly occasion
mutual hatred and provide pretexts for wars'. However, this
tendency should be countered not by weakening the various
energies but by moving towards a system of 'the most vigor-
ous' civil 'competition'. Risks will remain, but they must be
faced for the sake of freedom. What is not acceptable is a

despotism of peace, which would become 'the graveyard of freedom'.[4]

A unified world state is therefore the best option only in an abstract sense. The actually best solution is the theoretically second-best: namely, a world federation in which each state retains its sovereign rights (including the right to wage war) but undertakes to solve all conflicts through peaceful negotiation. But there does not exist a supreme authority capable of forcing states to observe the commitments they have freely made. It would remain as true as before that no superior instance could prevent an individual state from violating commonly agreed rules. Only an alliance operating at the same level would be able to compel respect for the rules, if necessary through force.

The conclusion to be drawn from Kant's reflections is that there will be no homogeneous and peaceful political universe. Politically speaking, the world remains a 'multiverse'. The basic 'natural condition' of hostility among states cannot be overcome but at best only regulated. That sounds pretty modest, but there is more to it than meets the eye. To bank on rules and regulations is to opt against a vertically established superpower and for the horizontally negotiated rule of law and right among states and nations. This rule of law, in place of world government, relies on recognition from shifting alliances of states and organizations as well as from a critical public. It is an insecure foundation, but there is no security in human affairs. The vision of a world state is delusive in this respect; it imagines the world to be beyond historical change.

In present-day conditions – when the US global power believes that it should ignore the international rule of law, not out of self-interest but with explicit reference to a policy of world peace – it is worth considering Kant's point that the 'duplicity of politics' makes 'respect for the law' more valuable than the so-called 'philanthropy' often invoked by those who do not wish their hands to be tied by the test of legality. A superior power, Kant writes, usually finds it advisable 'not to enter into any contract' but instead 'to reduce all duties to mere acts of goodwill'. This 'subterfuge of a secretive system of politics' must be exposed to public criticism, in the name of the principle that one should first comply with the legal rules and prohibitions before (as he smugly puts it) one enjoys 'the sweet sense of having acted justly'.⁵ When applied to a current problem, this means that general security requirements are not enough to vindicate attacks on so-called 'rogue states'; such attacks must justify themselves under international law. Even 'well-meaning' might does not come before right.

Kant argues for a rigorous formalism. Only in the law, only in the existence of a possibly good and well-meaning global power, does he see a chance (obviously no more than a chance) for world peace. There will never be a firm guarantee. For there will have to be powers – shifting powers and alliances of powers – that are prepared to defend the law, and therefore to do nothing less than place themselves under the power of the law. That is no small matter; perhaps it is too much for a political body with power interests of its own in the historical arena. And whether such law-protecting powers actually exist in history is partly a question of chance and circumstance.

Kant was enough of a realist to be prepared for uncertainties. This is also necessary if we look more closely at the social developments which, he thought, might improve the chances for global peace. There are three such trends favouring peace, although they are more limited than Kant hoped.

The first is the development of democracy: 'If . . . the consent of the citizens is required to decide whether or not war is to be declared, it is very natural that they will have great hesitation in embarking on so dangerous an enterprise. For this would mean calling down upon themselves all the miseries of war.'⁶ In the twentieth century, it was indeed not the democratic states that initiated wars. But the idea of the nation, itself a product of the democratic age, made possible undreamed-of mass mobilizations in the service of war. It was shown that nations readily consent to war in the belief that they are right – even when it is only the right of the strongest and a victorious clash of arms is to be expected.

The second trend is the civilizing power of world trade, on which Kant pinned so many of his hopes: 'The spirit of commerce sooner or later takes hold of every people, and it cannot exist side by side with war. And of all the powers (or means) at the disposal of the power of the state, financial power can probably be relied on most. Thus states find themselves compelled to promote the noble cause of peace.'⁷ Kant could have had no idea where an imperialism driven by economics would lead, or which new energies and motives would be brought to war by capitalist competition.

The third trend is the growing importance of the public, or the principle of publicity. Kant thought that, if political affairs are talked about in public, then war too must ultimately be defended in the arena of argument. Publicity places war under pressure to justify itself. What Kant could scarcely have imagined is that, in the age of the mass media, words and images would also decisively contribute to the mobilization for war and not always display the beneficent effect of 'discursive enlightenment'.

The constitution of politics out of democracy, market and publicity is for Kant a work of art; it provides the framework in which the individual can be a good citizen without first being reformed into a good person. Kant was so convinced of the cunning rationality of this system that he even asserted: 'The problem of setting up a state can be solved even by a nation of devils (so long as they possess understanding).'[8] It can be solved if the 'antagonism of hostile attitudes' *within a nation* is so ordered that they cancel one another out and lead to a condition of peace. But the devils must have 'understanding'. What does this mean exactly? It is the understanding associated with self-preservation. People must behave rationally and predictably from the point of view of self-preservation. Only then, as Kant was aware, will things work out.

The idea that wild self-preservation leads to hostile antagonism, whereas sensibly regulated self-preservation can help to bring about a condition of peace, does not take into account the fact that hostile antagonism can have other motives than unbridled self-assertion. One such motive, as we have seen, is the thymotic passion, the desire to be different and superior. Hostile antagonisms will not cease unless this tendency too is subjected to civilizing discipline. Therefore Kant undertakes an impressive attempt – it is the practical core of his philosophy – to found a *reason* that has in a sense absorbed the thymotic passion. This reason, unlike understanding, does not appeal to self-preservation but it does also appeal to pride: it may perhaps demand the sacrifice of one's own person, if human duty makes that necessary. In so far as it serves dignity rather than utility, this reason is actually thymotic in nature.

The kernel of Kant's argument is a belief in this proud, courageous, thymotic reason. The guiding image is that of a man who fits into his community by virtue of a capacity for peace, not because he is weak and conformist but because he is able to master himself, to control his egoism. For Kant such reason is universal. It is the organ that helps the individual to understand himself as a member not only of a nation and a state but of humanity. Reason removes limits. The individual who respects and listens to his reason thereby respects and discovers the humanity within him – humanity not in the sense of a statistical quantity but as an expression of the rational dignity of the individual. Whoever honours the 'humanity' within him overcomes the naked interest of self-preservation and becomes capable of solidarity. It is this reason that, for Kant, makes man a citizen of the world. It is the direct path from the I to the We. In the light of reason, the age of reconciliation can finally dawn.

But now reason begins to dream of reason. One of its dreams is that the adversaries of reason might lose some of their strength, and in particular that the dogmatic religions – which for Kant are the main adversary of reason – might cool down into their common moral core (what Hans Küng today calls the 'Global Ethic'). It is a hope that religions might stop justifying all manner of cruelties in the name of an exclusive god.

In reality, however, there is again a threat that resurgent religions will bring hostile antagonism in their path. Christian Europe has previous experience of that, in the shape of wars and untold horrors committed in God's name. Europe was able to overcome religious bestiality only by separating religion and politics: that is, by politically neutralizing religions and forcing them to coexist with one another in the public space. Today, we again have cause to remember that civilizing victory which followed the bloody lessons of the religious wars of the sixteenth and seventeenth centuries. The point is not to combat faith and religion but to adopt the basic principle that whoever loves God more than men makes him into an idol in whose name terrible acts of cruelty can be committed. The Enlightenment critique of religion, such as we find in Kant, is directed against that kind of misanthropic love of God. It is still relevant today. For, whereas in the West nihilism and decadence are in the ascendant, elsewhere it is religious enmity that is making headway. Dostoevsky once wrote: 'If God is dead everything is permitted.' That may be correct. But history teaches us that people already permitted themselves everything by invoking the name of God. There are gods that incite people to the worst crimes.

Everything indicates that it is those gods who will find support among the growing number of losers from globalization. Loss of tradition, deracination and the practical nihilism of consumer culture are the breeding-ground for the deliberate and militant re-enchantment of a perverted religion. In the first half of the twentieth century, the totalitarian ideologies of socialism and fascism played the role of perverted religion in revolt against the impositions of a secular pluralist modernity. Today it is Islamic fundamentalism that continues this totalitarian tradition. One does not need to give a precise definition of genuine religion in order to recognize a perverted religion for what it is: bestiality and stupidity are already conclusive evidence.

The world-views of perverted religion claim to know the true essence of nature and history, the innermost secrets of the world. They wish to comprehend the totality and reach out for the whole man. They give him the security of a fortress, complete with observation slits and crenels. They count on people's fear of open lives, of the risk of human freedom that always involves insecurity, uncertainty and standing on one's own two feet. They seek to free people of their difficult freedom to be individuals, by incorporating them into a collective where they can feel a sense of belonging in hostile opposition to those who do not belong. This opposition is elemental: such a sense of belonging, when looked at closely, is nothing other than the boundary between friends and enemies. Perverted religion lifts the burden of freedom, which always includes feelings of estrangement and solitude.

There can be no doubt that Western culture, which rests on the principle of freedom and the separation of religion and politics, is a 'cold' project. Here, religious truths have by now learned a degree of modesty, so that they can endure the insult of being treated on the open market as mere 'opinions' or 'convictions'. A papal encyclical has to compete with designer counselling, and the Bible with other esoterica. We in the West have entered the age of secular polytheism. In pluralist society there are many gods, many guides to the world, many religious and semi-religious definitions of the meaning of life. The One God who used to guarantee the spiritual coherence of Western society has split into a multiplicity of household gods. While the main churches stand empty, the demand grows for religious hobby venues. But where the main churches engage in public activity, they usually do so as tax-funded administrators of the social residue of spirituality. They legitimate themselves with a view to socially desirable effects. One aspect of the West's 'cold' project is that religious institutions have sought to occupy a new social niche through the provision of edifying goods and services such as social work, a meaning in life and public holiday events.

All this must appear contemptible to the 'heated' funda-mentalist desire for meaning. The West has found happiness in disenchantment and estrangement, and is therefore under attack from a religious fanaticism that has again set its heart on the totality. The sinister 'clash of civilizations' has already broken out – in the Western world and on its borders. The conflict again makes it clear which spiritual prerequi-site has to be fulfilled before the Western democratic way of life becomes even a possibility: that is, the principle of the separation of powers must be internalized. The separation of powers is here meant in the broader sense in which it not only regulates the political game but has implications even for the understanding of 'truth'. For the force of truths, especially when they appear in religious garb, is thereby restricted. Separation of powers in the spiritual domain signi-fies de-escalation with regard to truth claims. We have learned that truth too is divisible, that it is divided among individuals, that a single indivisible truth is not within human reach. The idea of the separation of powers is therefore bound up with the principle of tolerance. I will fight against your opinion, but I will also fight for your right to express it – that is how Voltaire famously defined tolerance.

The separation of powers not only regulates how people interact with one another but reaches beyond that into the individual self. The individual becomes accustomed to living in different spheres of value. Politics, science, economics, art, religion and the family are different areas in which people think, feel and act differently. Everyone inhabits several worlds at once. The inner separation of powers also means that people produce within themselves a balance among the various powers in which they are entangled. Thus, the principle applies not only to the organization of the polity but also to its individual members. Let us not deceive ourselves, however: the complex system of the separation of powers demands too much of individuals, and they would gladly lead an integrated life without having to fight out, inside themselves, the conflict among different spheres of value. For this reason, it is at the contested fault-lines that we first realize the fragility of the Western way of life, which perhaps requires too much to count as a global paradigm of socialization.

For us beneficiaries, however, it is still worth defending the Western model – even if, or especially if, it is agreed that democracy plus freedom of belief and speech plus separation of powers plus separation of religion and politics add up to a rather rare specimen in human history, and that there is little to suggest it might ever triumph all over the world.

6
The Global and the Other Totality

It is not only in relation to world peace that global thinking finds itself caught in uncertainty; the same is true of a whole maze of other problems, although globalist ideology and its patent remedies would gladly turn a blind eye to them. Globality forms a complex context, and action within it usually brings consequences other than those originally intended. It is true that this has always been the case, but today we know more about it and can no longer remain blissfully ignorant. For the horizon of global problems now imposes itself even on our everyday consciousness, with the result that there are more and more occasions on which we feel our lack of power. Our particular lifeworld is no longer a sheltered area. Almost every change in our immediate surroundings – in our work, food, transport, media use or health matters – can be understood as the last link in a causal chain that we cannot see as a whole, but about which we know so much that it stretches far back into the highly complex global web.

In the past, special elites were responsible for the wider whole – first priests and philosophers, then also politicians. Those who had demonstrative dealings with the whole were regarded as exceptional people. In the politically expansive world of the great empires, there were growing demands on these elites: they had not only to form a picture of the whole but to make it the object of concrete political activity. The

Roman Empire was the first to witness politics on the grand scale: *urbi et orbi.*

One aspect of modern globalization is that global thinking becomes more democratic and concerns general public awareness. It is therefore only logical that there should also be an ideological globalism, which is the present expression of a politicization that has marked large parts of modern consciousness. It recalls the major turning point at the end of the eighteenth century.

Under absolutism, politics was a monopoly of the monarchical state; the monarch was absolute since he or she held undivided political power. Society was free of the absolutist state in two senses: it did not (with some exceptions) seek political forms of expression, and the state did not draw it into politics. The late-absolutist monopoly of politics was tersely expressed in a decree issued in 1767 by the Prussian government: 'A private individual is not authorized to pass public or critical judgement . . . on the actions, procedures, laws, regulations or orders of the sovereigns and courts, or to make known and publicize in print information that is given to him about them.' The 'private individuals' barred from politics responded by making a new evaluation of their inner life, morality and culture. They brought 'man' to bear against politics, using a definition that made man superior to anything merely political. Schiller, for example, taught in 1784 that the theatre had to offer the public a stage for the emotion of 'being human', and that it had the function of a law court. The jurisdiction of the stage in placing the political process on trial began 'where the influence of civil laws ends'. The spirit of true 'humanity' was thus turned against a political world that was experienced as too narrow. The jurisdiction of the stage thus had to be both strict and self-assured: 'there are a thousand vices unnoticed by human justice, but condemned by the stage; so, also, a thousand virtues overlooked by man's laws are honoured on the stage.'[1]

The historical sequel is well known: at the time of the French Revolution, the ethical and literary public turned into a political public that demanded openness from absolutism as a matter of principle. Society thereby shattered the absolutist monopoly of politics and recovered it for itself, with the result that the sphere of the political underwent a complete transformation. Politics became an affair of man as a whole and of the masses. In this sense it became 'total'. The *levée en masse*, the general force of the revolutionary armies that flooded Europe, not only put an end to the old showpiece wars and mercenary formations but fundamentally altered the relationship of ordinary people to history. Whereas history had formerly engulfed people in the manner of a natural disaster, it was now understood as a process into which everyone could be drawn as a potential activist. Everyone was permitted to lend a hand in the making of history. When Napoleon declared that 'Politics is destiny', he was saying not only that it decides everything but that it is a destiny we can grasp with both hands. This politics, which reached out to the whole man and freed the masses for history, occupied the place that had previously been claimed by the numinous, the sublime and the transcendent. Politics became an affair of the heart. But this totalization of politics also meant that anything that could not be translated into politics counted as irrelevant.

The epochal thrust of politicization around the turn of the nineteenth century should be thought of as an early form of globalism. It reconfigured the relationship between the individual and the whole, and redirected the question of the meaning of life away from religion and towards politics. A secular impetus converted the so-called 'ultimate questions' into social and political issues. Robespierre staged a divine service to political reason, and the Prussian wars of liberation in 1813 first brought into circulation the prayer-books of a patriotism that was about to turn into nationalism.

Politicization marked the first dramatic narrowing in the perception of the whole. The mid-nineteenth century brought the second, in the shape of economization. Now the power over destiny and interpretation was claimed by an economism for which money was the currency of values and commodities were the truth of the world. Money and trade really did bind everything to everything else, as they penetrated into the last hidden corner of both society and the individual. If the money for such diverse things as the Bible, schnapps and sexual intercourse created a common expression of value, then we may see in it a link to Nicholas of Cusa's concept of God as the *coincidentia oppositorum*, the point where all oppositions merge into one. Money, as the exchange-equivalent of all values, rose godlike above the manifold of the world of appearance. It became a centre where the different and the opposite found their commonality. The circulatory force of money had outstripped the spirit of which it used to be said that it bloweth where it listeth . . .

Politicization and economism, those two reductions in the perception of the whole, converge in contemporary globalism. It is no wonder, then, that people can nowadays think of globality only with a sense of unease and constriction. Economism awakens the idea of an ultimately monotone universe consisting of workhouse, market, financial flows and trade in goods; while politicization restricts thought to the dimension of strategy and counter-strategy. In the political way of seeing, globality is impoverished to such an extent that it becomes a mere object for calculations of power or impotence. In other words, even thought lands in a global trap when it tries to reach out to the whole. It circles monotonously around two basic questions, some asking how we can control the global, others how we can rescue it. The same applies to mental pictures as to thoughts: they remain captive to endlessly viewed TV images of poverty and wealth, violence and pleasure, wilderness and metropolis. For the media-supported global consciousness, there is no sense of wide open spaces. Most things appear familiar without being really known, unsurprising yet still unsettling. A world is shown that is 'afflicted with similarity' (Adorno). The global has ceased to entice from afar. The constricting nature of globalist perceptions appears most clearly in contrast to the historically older, cosmopolitan way of relating to the totality. Ever since it has existed as an articulate experience, cosmopolitanism has had the function of a stipulation of openness.

The Cynic philosopher Diogenes of Sinope, the opponent of the great Plato, brought the term *kosmopolites* (inhabitant of the cosmos) into circulation. To the reproachful question why he had departed from the morals and thinking of the city, he answered that he was an inhabitant not of a city but of the cosmos. The cosmos was thus the great whole to which people really belonged, and where they found refuge from the restrictive laws and regulations of the *polis*. Those laws become relative within the perspective of a higher order. If cities, peoples and nations are thought of as communities based on arousal, then the double sense of cosmopolitanism becomes apparent. It seeks to sober people up, but also offers a special kind of intoxication. The sobering power of cosmopolitanism comes from its reminder of certain basic features of the human condition. Aristippus tells us that the distance to Hades is the same from everywhere; the cosmos reminds us all of our mortality. Only in the city are we tempted to overestimate our own significance. But there is also the intoxicating side of cosmopolitanism. For the cosmos has a frontier not only with Hades but also with Heaven, where God rises above all the gods of particular tribes and cities. The cosmopolitan transcends the sphere governed by human law in both of two directions: towards an awareness that Hades is close to us all, and towards an awareness of a 'community composed of men and God', as Epictetus puts it.

There were times when even this cosmos was too narrow for many. Gnosticism, the great spiritual adversary of early Christianity, condemned the whole earth as a botched creation, a prison for the soul. And it was little wonder, since the earth had meanwhile become coextensive with the Roman Empire. The known world had, as it were, turned into a single city, and so the cosmopolitan removal of boundaries rose beyond the known cosmos into the non-cosmic realm. Christianity, on the other hand, was an inspired attempt to build a spiritual empire into the Roman world empire. The borders of the two might coincide on earth, but the spiritual empire of the Christians was open to the sphere above. There was a clear preoccupation with spaciousness. The important element, then as now, was spaciousness that was or could be animated.

In modern times cosmopolitanism did not lose its function as a stipulation of openness. When Dante wanted to escape the strife among the states of northern Italy, and to avoid being drawn into the passions aroused by civil war, he too described himself as a cosmopolitan poet whose homeland was the world.

A cosmopolitan who loosens his social and political ties by appealing to the wider circle of humanity inevitably brings opponents into the arena. They reproach him with irresponsible escape into a realm of ideas where it is possible to dream but not live, to wax ecstatic but not to act. Man, they argue, must become a reality in a particular community – a tribe, a people, a nation – or else have no reality. This criticism is not directed against the principle of cosmopolitanism, but only against its escapist, perhaps even hypocritical, tendencies. For Fichte, the task was to think as a cosmopolitan while acting as a patriot; the cosmic and the universal-human must not float above and around us, but acquire definite form in our distinctive *patria*. Fichte therefore advocated a cosmopolitanism that would not be afraid to sink roots in a 'closed trading state'.

This was still friendly criticism. As we know, however, in the age of nationalist and socialist totalitarianism, cosmopolitanism was explicitly branded as hostile and threatening. In the totalitarian 'people's prisons', the cosmopolitan inevitably appeared as an unreliable type, or at worst as a traitor. This was impressive confirmation that cosmopolitanism served as a stipulation of openness. Today's globalism no longer has such a function. With this in mind, let us once again summarize the various forms of globalism.

Neoliberal ideology portrays capital as cosmopolitan, if 'my fatherland is where my business prospers' can count as a cosmopolitan statement. Actually, capital has many fatherlands: it is always at home where it can be put to profitable use. In this sense global players who live on the routes of their capital – e-mailing, faxing and jetting their way around the world – are cosmopolitan. Globalism, as the ideology of global players, also serves as a stipulation of openness, but not in the same way as cosmopolitanism. It demands that markets be opened for investment capital, but the principle of the closed trading state keeps the products and services of the Third World from circulating. Globalism qua anti-nationalism (or multiculturalism) turns out, especially in Germany, to be not an opening but a mere manoeuvre to shed the burden of responsibility. In flight from their own history, people seek refuge in the totality.

Finally, with regard to ecological-ecumenical globalism, we may say that its main investment is not in a cosmopolitan opening-up of space but in a global claustrophobia. In any event, globalism makes spaces narrow. When it is genuinely sensitive, moral and responsible, it piles up problems that can drive one to despair.

Globalism is a symptom of excessive demands. Evidently no one can endure globalization – hence the way in which people wall themselves into various ideologies (neoliberalism, multiculturalism, and so on); hence too the flight into fantasies of collapse and salvation. Of course, there is also a more sober, politically sophisticated way of dealing with problems of globalization, which is driven by a sense of justice. The critics of globalization in the French Attac movement, for example, do not wallow in visions of global collapse, but circulate analyses, uncover contradictions and scandals, and suggest pragmatic forms of resistance. Even within the power apparatuses of mainstream politics there are some signs of rethinking. Nevertheless, or perhaps for that very reason, the global domain has become the arena for economics, the media, politics, strategies and counter-strategies. It is no longer the whole envisioned by theology, metaphysics, universalism and cosmopolitanism; it is a whole that has become the object of economic, technological and political processing. This explains the peculiar sense of constriction on a global scale. Everything, even the bad news, seems somehow familiar. Global imperatives resound from every part of the world. Each new item of information also conveys a sense of impotence. Globality appears as a systemic link, operating in the end without a subject yet with such force that it seems almost obscene to recall the significance of the individual.

But that is what needs to be done. We must turn the stage around and realize that, while our head is in the world, the world is also in our head. To be sure, individuals are nothing without the whole of which they are part. But the converse is also true. There would be no whole at all if it was not reflected in our heads, in each person's head. Each individual is the stage where the world makes its entrance, where it can make its appearance. The world will be meaningful or empty according to whether the individual is bright or dull. To shape globalization is therefore still a task that can be handled only if the other major task is not neglected: the task of shaping individuality itself. For the individual is also a whole where heaven and earth touch.

7

The Individual and the Immune System

Let us then return from the global sphere to people as individuals, within the limits of their lifeworlds and lifespans. Here everyone must define for themselves the relationship between what they can conceive and what they can actually live. We first become individuals by giving ourselves a shape or outline, and therefore by drawing boundaries. This is a difficult task, and a whole world of ideas – the world of individualism – has been developed to cope with it. Individualism is an important achievement of Europe's political and philosophical culture. It is based on the idea that human diversity is not only a fact but an asset worth preserving; that not only do individuals exist but individuality should exist. This normative decision is bound up with the argument that the state and social life must be organized in such a way that people can fully develop their individuality, without impeding one another in the process. This conception of plurality as something we ought to encourage and protect gives rise to the other Enlightenment norms that make up European modernity: freedom of opinion and conscience, toleration, justice and freedom from bodily harm.

Of course, the defence and development of individuality has a long prehistory that we cannot relate here. The legacy of antiquity and especially Christianity has played a role in this. A religion whose god addresses people with the familiar 'thou' creates favourable conditions for individuality to be taken seriously. Yet individualism is a product of secularization, for it presupposes that religious statements have lost their validity with regard to the meaning and purpose of the whole. Individualism invests meaning in individuals, and no longer in such totalities as God, humanity, nation or state. The meaning of the whole is that it makes a rich individuality possible. But individuals are rich only if they discover and develop their own wealth. The individual – every individual – constitutes the core of meaning for the whole.

Crucial here is the belief in human self-malleability, in the possibility of making much of ourselves. In a sense, the ban on figurative representation has been shifted to the individual: thou shalt not make to thyself any graven image, either of God or of human beings. In educating themselves for individuality, people break down any fixed image and should therefore not be tied down to any. Every individual, as Wilhelm von Humboldt taught, enriches the concept of humanity by producing a new characteristic form. Such individualism, while wagering on freedom as its essential prerequisite, leaves it open where freedom should lead – not because nothing more could be said on the matter, but because one would have to say too much, to explain how it is that there are individuals in the world. Freedom is creative, and individualism seeks to preserve the conditions for it. It does not normatively tie this self-creative freedom to certain results or achievements, but defends (also normatively) the presuppositions of freedom. This idea, which is at the basis of liberal democracy, does not want people to become absorbed into a whole; it seeks to ascertain, both theoretically and practically, the opportunities for individuals themselves to become something whole. A fundamental precondition of individuality here comes into view: namely, that power to define oneself which, as it were, provides immunity from powerful lures and unlimited horizons.

Goethe, who knew exactly what is required for the education of an individual, wrote in *Wilhelm Meisters Lehrjahre*:

Man is born to fit into a limited situation; he can understand simple, close and definite purposes, and he gets used to employing the means which are near at hand; but as soon as he goes any distance, he knows neither what he will nor what he should be doing, and it is all one whether he is distracted [or 'scattered' – *zerstreut*] by the large number of objects or whether he is put out by their greatness and dignity. It is always a misfortune when a man is induced to strive for something with which he cannot associate himself through some regular spontaneous activity.[1]

Here, as so often, Goethe is right on target. There is a range of our senses and a range of activity carrying individual responsibility, a circle of perception and a circle of action. Lures – to simplify greatly – must be somehow removed from our path, initially by means of an action-response. Action is the appropriate reaction to the stimulus of a lure. Thus, both the circle of sense stimuli and the circle of their removal through action are originally coordinated with each other – but not sufficiently well coordinated. Here too we are semi-finished products. We have to develop for ourselves a system to filter out stimuli to which we cannot adequately react or to which we do not need to react. Our senses are perhaps too open. Our immune system is not adequate in this regard. This too is part of the work on our second nature: the development through culture of a filter and immune defence system.

We have until now criminally neglected this task, at a time when a global information community has been taking shape through the media. The emergence of this community means that the quantity of stimuli and information dramatically exceeds our range of possible action. The circle of our senses, artificially enlarged by media prostheses, has completely detached itself from our circle of action; there is no longer any adequate way of responding to it, and therefore of removing the excitation and converting it into action. While the scope for individual action dwindles away, the pitiless logic of the media market increases the inflow of information and images. This has to be the case, because those who offer the stimuli are competing for the scarce commodity of 'public attention'. But the public, grown accustomed to and then dependent upon sensations, demands an ever higher (or anyway ever new) dose of excitement – instead of an action output, an arousal input.

What happens with stimuli that do not elicit an output response in action? Hardened receivers of stimuli have perhaps rendered them harmless through the very process of hardening. But even among them the stimuli leave traces. If they are not filtered out, they are deposited somewhere inside us, in a new and troublesome area of the unconscious, flexible enough to arouse us while being only loosely associated with their 'objects'. Individuals become 'scattered', as Goethe put it. But it is an agitated scattering – as after an explosion. We should think of this as follows: everyone is at first compressed in their work and other daily activity; when this concentrated pressure subsides, in so-called free time, those freed from the pressure fall apart and pounce on a thousand images of events that do not actually concern them – though in reality it is the images that crowd in on them. In any event, the media consumer experiences the global world as the scene of his on-the-spot agitation, a kind of agitation that is meant to last and is continually searching for new 'occasional causes'. Knocked into us through the media, globalization encourages latent hysteria and panic states. This also produces a kind of long-distance political moralism, a 'tele-ethics' for the television age. One need only think of the examples of former Yugoslavia and post-9/11 Afghanistan.

Let us consider the methods of warfare in the Balkans. They had an effect: the civil war was contained, and the warmonger Milosevic was overthrown. It is a matter of dispute whether there had to be so many casualties and so much destruction among the civilian population. But the bombing campaign from the air accurately reflected the mood in the West that televised images had helped to generate. Something had to be done: the morality of the Western superego demanded it. And it suited that superego (which had to be satisfied almost more than the people directly affected) that things should be done from a great height. Warfare with little ground contact is the real-world counterpart to an armchair moral commitment in which risk-free assistance is given at the level of the imagination. Wars conducted under the spotlight of television produce a new type of media 'freeloader'. Bombers avoided the risk of being shot down by flying too high to hit targets accurately, with the result that their bombs hit innocent people and civilian targets on a large scale. It was as if one were to deal with a bank robbery by blowing up the bank instead of doing away with the robbers.

This peculiar form of aid, which shifts most of the risk on to innocent civilians ('collateral damage'), poses a problem of morality typical of the age of media warfare. Terrible news items force themselves upon us, making us feel a moral obligation to act. Faraway places move closer, but any action is supposed to remain action from afar or from a great height. Bombs are rained down from the sky, but care is taken to avoid ground fighting and all the risks that come from entanglement in local conditions.

In Kosovo, ground contact was a highly explosive question for morality as well as for military tactics. More generally, however, the whole issue is symptomatic of a kind of global mobility that shuns close ties with any of the complex lifeworlds on the ground. Locations resemble shifting dunes. Or, to use a different image, weak road-holding is one of the basic features of globalization.

As far as Afghanistan is concerned, the pictures of 9/11 that were burned into people's minds in the West added a sense that they themselves were under threat. It was a traumatic experience to discover just how vulnerable was our complex high-tech civilization. Used to thinking in terms of 'systems', where the individual as such scarcely counted, people suddenly realized what a huge impact a few totally determined individuals could have. Skilful use of the infrastructure against 'the system' made it possible for a handful of men brandishing carpet knives to shake the world with a blazing inferno. Now the individual again counts for something: such is the lesson that terrorism teaches, in its appalling way. The measures really necessary to fight crime then no longer appear spectacular enough to minds in which the media have sown panic and hysteria. And so it must be a 'war' that is (part genuinely, part rhetorically) waged against terrorism, even though non-state violence and molecular civil war have little to do with the traditional forms of interstate warfare.

Terrorism operates at a symbolic as well as a material level: there are terrorist actions but also, equally important, the dissemination of terrible news. The media thereby become unwilling accomplices. Some sow terror in the expectation that others will spread it. Of course, the messenger is not responsible for the bad news, but that does not alter the fact that the use of media messages is part and parcel of modern terrorism. This poses a real dilemma: one would have to prohibit the spread of terrifying news, by analogy with medical practices to block the transmission of diseases, but that would run contrary to the duty to inform the public. The media revolution too devours its children.

The distance-closing principle applies to both examples: to tele-ethics in the television age, and to the involuntary complicity between terrorism and the media. The global system is becoming a simultaneous whole: what happens over there can be immediately seen here in real time; here becomes somewhere else and eventually everywhere. The highly graduated system of perceptual horizons radiating out from individual bodies has been breaking down, as a homogeneous space and a new simultaneity have come into existence. Faraway places burden us with a deceptive closeness, and the simultaneity from which distance used to protect us forces itself into what we think of as our own time. In the past, when something happened in a remote place, it was already long over when people elsewhere received news of it. The event had enough time to associate itself with various imaginings and interpretations, so that it had already been processed before it arrived. Faraway events never lost their faraway character and, because of the long communication time, acquired some features of the legendary, the symbolic and the spiritual. Indeed this was scarcely surprising, since the faraway character was communicated in the medium of language, and linguistic representation preserves the remoteness of what is represented. But that which Walter Benjamin once defined as the 'aura' – 'the unique phenomenon of a distance, however close it may be'[2] – has since been dissipated. Everything today is seen in close-up, no matter how far away it is. In tele-communication, and especially in tele-vision, we see the consummation of a trend that began with technically more robust forms of the overcoming of space. Sensitive observers already had an intimation of this in the early nineteenth century. 'Through the railway', wrote Heinrich Heine, 'space is suppressed and all that is left is time . . . It feels as if the mountains and forests of all lands had advanced on Paris. I can already smell the fragrance of German limes; the North Sea breaks before my door.'[3]

Despite the new technology for the virtual suppression of space – television, telephony, data transmission – physical travel has also been on the increase. Whether for business or to pass the time, people roam around incessantly. As we know, this mobile life contributes a thing or two to the devastation of space, as if the time so passed were taking its revenge on space. New roads, railways and airports eat up landscapes so that great distances can be overcome in the shortest possible time. The high speed actually forces the building of routes along which people can move without obstruction or troublesome involvement. The sense of crossing space begins to fade. There is a kind of tunnel between departure and arrival; the time spent there is supposed to pass as on a flight, preferably without any experience that has anything to do with the fact of travelling. Indeed, the publicity claims that on a good road, in a good car or on a fast train one will arrive the same as when one left: fresh, unruffled, relaxed. The effort of getting to a distant place is supposedly becoming a thing of the past.

But we experience something only if we approach it. To be somewhere too fast is to be nowhere. It is said of Australian aborigines that, after a long trip on foot, they sit for a few hours before their destination so that their souls have time to catch up. When travelling was an experience, one used to arrive a different person. Today, those who arrive the same as when they left will tend to see all destinations as the same: global mobility makes spaces uniform; and the flows of goods, capital and information make conditions the same wherever they end up. But that which is not touched by them does not remain untouched. It becomes something excluded or neglected, a marginal remnant that is part of the whole by not being part of it. One way or another, the local becomes 'glocal' – as they say nowadays.

The closing of distance is thus a basic feature of the globalized world. If proximity and distance become confused in the artificially enlarged perceptual horizon, this restricts the ability of individual space–time coordinates to provide proper bearings. New routines therefore emerge for trouble-free format changes, allowing potentially virtuoso performances with proximity and distance, individuality and totality. Man, a creature prepared for worries, now learns to worry about the future of the planet. Various global threats – melting of the polar ice, genetic manipulation, overkill capacities, disappearance of the rainforest, the spread of Aids, the ozone hole, the pension-funding gap – become part of the scenery in our immediate lifeworld. Wherever we go, we land in situations which make us uneasy that the extent of our familiarity with the global is becoming dramatically out of step with our capacity for action. The time is gone when the range of possible action was protected by lack of knowledge, when action was associated with a local area for which it was still possible to take responsibility. A life of improvisation in one's immediate surroundings has lost its innocence.

Since, as we know, everything is connected with everything else yet knowledge of this remains indistinct, individuals suddenly find themselves caught in a web of new imperatives and appeals. What are you doing about the ozone hole, about worldwide terrorism, about child labour in East Timor, about the repression of the Ogushen? How should this be handled? The superego as Freud imagined it is perhaps harmless in comparison with the superego that implants a sense of responsibility for the global future. True, we are again allowed to speak about trees, but it has to be a conversation about the death of the forest if it is to have any place in an ethics of the future.

No one can endure that kind of thing for ever. So, what happens is what always happens in such cases: people get involved in the splitting game. There is the public, global future and the private future. Worries about one's own provision in old age peacefully coexist with passionate sermons about the impending apocalypse. With each love story or home redecoration, each new job or book project, we branch away from a future of almost zero opportunity into a little private future that will allow us to continue muddling through, only moderately held back by the public ethic of disaster avoidance. To escape the panic that comes from constant overstretching, we try the shrewd path of 'as if' living. It becomes possible to lead a split existence in relation to the future of the planet. Besides, there is always the useful device of a division of labour.

The unreasonable demands of the superego have always been fended off with the help of delegates. There have to be superego specialists – priests and philosophers in the past, for example – who enable me not to do everything personally that the superego requires of me. The clouds on the global horizon too can be endured only through such a division of labour. The problem is that we have little reason to trust one inch the new global specialists. For the players in the arena where global negotiations really have an impact are mostly the ones with strong economic and political interests of their own. It would indeed be nice if people thought globally and acted locally. Usually, however, the opposite is the case: global players pursue the narrowest local interests – but with a global reach.

8

Jungle and Clearing

The globe used to be the challenging mystery par excellence, but the global present has ceased to be a mystery. The reason for this is that increasingly we move in a sign world of our own making. It is a technologically supported, globally networked civilization in which man has increasingly to deal only with himself – that is, with traces of his own activity. We are faced with the problem that our lifeworld is almost entirely boxed into second nature – that is, into an artificial world. This gives rise to a quite special kind of boredom. Human life becomes tautological if it encounters only traces of its own activity.

Metaphorically speaking, globalization is a planetary clearing of first nature in order that second nature can be put in its place; civilizations have been hewn out of the forest as a kind of clearing. This metaphor of forest and clearing allows the history of our civilization to be told as a short story. It was Giambattista Vico, the founder of modern historical theory, who first attempted this in his philosophical fantasy, *The New Science*, when he made history begin with the 'giants' who cleared the first forests. Those giants lived in forests; they knew no sky because the leaf canopy blocked it out. A terrible thunderstorm tore open a piece of forest or *lucus*, which, Vico writes, means both eye and clearing. For the first time the giants could see the sky and feel that it saw them. They were no longer alone. Then, with the giants in the open, there began the drama of seeing and visibility, hearing and being heard. There was no holding back as a communication society, the first human network, took shape in the clearing. Vico writes: 'First the woods, then cultivated fields and huts, next little houses and villages, thence cities, finally academies and philosophers: this is the order of all progress from the first origins.'[1] Vico leaves no doubt that, since there have been academies, the clearings have begun to change into wasteland. It was the year 1744 when his *The New Science* first appeared.

The clearing, Heidegger tells us, is a place of truth, of de-concealment. Must one not first create space to recognize something? Must one not fell trees to be able to recognize them as trees? Does the triumph of truth mean that the wasteland grows? Do the forests not always remain a dark torment for the cultures hewn out as clearings within them?

Here is another story, the oldest of them all: the Gilgamesh epic that was recorded by the Sumerians two thousand years before the birth of Christ. Gilgamesh, the fabled king of Uruk, is the first enemy of the forest. He would like to conquer the 'Land of Cedars', slay the forest demon and fell the trees. He has looked over the walls of his city and seen how dead bodies drift down the river. Gilgamesh rebels against mortality; he builds a new city, with walls to keep out the rampant and rotting vegetation: 'I will go to the country where the cedar is felled. I will set up my name.'[2] In the end despair catches up with him: he sees his own corpse drifting among the severed trunks that he sends down river; he himself will fall in the clearing that he makes. With the dawn of civilization, this simple realization becomes known as 'wisdom'. It recalls the difficult art of living with the forests around and within us, and therefore with our status as creatures. It is a difficult art, because we are not only natural but also artificial. We transcend nature by making something of ourselves. We stand somewhere between god and beast, tree and bark. The epic of Gilgamesh puts it like this: 'Whoever is tallest among men cannot reach the heavens, and the greatest cannot encompass the earth.'[3] In Greek myth Pentheus, the hero of the city who rebels against the dark natural power of Dionysus, suffers a similar fate to that of Gilgamesh. He is torn limb from limb by the frenzied women in the forests of Dionysus, where he has gone in pursuit of the law. The people from the city can do no more than gather together his body parts strewn in the forest.

For the ancient Greeks there remained a tragic tension between the dark of the forest and the order of the city, between nature and civilization. Nor was this tension lost in the Christian era.

> Half way along the road we have to go,
> I found myself obscured in a great forest,
> Bewildered, and I knew I had lost my way.[4]

So begins Dante's *Divine Comedy*. The forest becomes Limbo. It is the wilderness of sin and animality. Only with divine assistance can human beings find a way to freedom. Therefore, the whole of history can also be told as a 'comedy'; the divine accompaniment even makes of it a divine comedy. This overcomes the ancient tragic collision between nature and city, wilderness and civilization. In the Christian promise, we shall be able to triumph over the forests within and around us.

While Dante wagers on God, Descartes wagers on a reason in which the god of geometry lives. In his *Discourse on Method*, Descartes tells the story of the path he found through doubt and despair to the truth. He first had to establish a few 'provisional' (geometrical and mathematical) principles, then follow them strictly and see what resulted. He modelled himself on travellers

> who, finding themselves lost in a forest, know that they ought not to wander first to one side and then to the other, nor, still less, to stop in one place, but understand that they should continue to walk as straight as they can in one direction, not diverging for any slight reason, even though it was possibly chance alone that first determined them in their choice. By this means if they do not go exactly where they wish, they will at least arrive somewhere at the end, where probably they will be better off than in the middle of the forest.[5]

We know all too well where Descartes's path led, since we still find ourselves where he arrived: in the cities *more geometrico*. These are different from the old winding cities that grew up without being designed by a central will – cities that Descartes shudders to compare with 'those which are regularly laid out

on a plain by a surveyor who is free to follow his own ideas.'6
It is from such vantage points that we look out today and see
forests disappearing in smoke, timber plantations and parks.
Something must have gone wrong. Similar doubts already
plagued the late eighteenth century. Rousseau sought the solitude of the forest in city parks and country estates. There he
dreamed of returning to a nature done up in the style of the
place where he was staying, an oasis of civilization, domesti-
cated, cultivated and innocuous. Quite different from a
wilderness. Rousseau longed for the romantic forest of the
noble savage, in contrast to the monsters living in civilization.

In the nineteenth century, forests became for some the sanc-
tuary of Goodness, Beauty and Truth, and for others the niche
of the mysterious and the uncanny. But they were still nostal-
gic, symbolic forests, a poetic-philosophical resource stretch-
ing from the tales of the German Romantics down to
Heidegger's *Holzwege*.

The whole is a history of the slow dismantling of the potential of myth, a history of disenchantment and clearance and
expanding wasteland. The world goes on the Net – and falls
into the net of uniform globalization. What disappears here
is the external equivalent of man's inner transcendence – a
transcendence without metaphysical constructions. In the
impenetrability of external nature, symbolized by 'the forest',
we, who are also nature, experience our own mystery. The
forests reflect back to us the alien quality that prevails even in
our relationship to ourselves. Whoever releases this alienness
into the wild remains in contact with the mystery of life. But
a word of caution: this mystery does not promise redemption.
We find no answer in 'the forest', only a puzzling echo of our
questions. Henry David Thoreau would learn this in 1845
when, on the anniversary of the American declaration of inde-
pendence, he went to live for a time in a hut that he had built
for himself in the forest. On the reasons for his forest journey,
he wrote:

> I went to the woods because I wished to live deliberately, to
> front only the essential facts of life, and see if I could not learn
> what it had to teach, and not, when I came to die, discover that
> I had not lived. . . . Let us settle ourselves, and work and
> wedge our feet downward through the mud and slush of

opinion, and prejudice, and tradition, and delusion, and appearance, that alluvion which covers the globe, through Paris and London, through New York and Boston and Concord, through church and state, through poetry and philosophy and religion, till we come to a hard bottom and rocks in place, which we can call reality.[7]

Thoreau looked in the forest for authentic reality. But he learned from life there that nature reminded him of his own estrangement. It did not remove the responsibility for his own existence. It did not bestow the salvation of prenatal security. Nature is not a mother: it leaves us our freedom, and therefore the experience of alienness and estrangement without which there can be no freedom. We discover ourselves in nature amid something absolute, which we do not possess and which refuses to take us into its care. After a year Thoreau left the forest. The nature that had shown itself to him remained indifferent to his dreams, but he returned to the city a changed man.

Thoreau had gone to the forest because he felt cramped in the concrete jungle. He anticipated a situation that we are approaching today: as the real forests slowly recede, they return in another dimension as a 'forest' of signs, information, interpretations and signals. Thoreau spoke of the 'mud and slush of opinion, and prejudice, and tradition, and delusion, and appearance'. But he learned that escape into the forest did not free him from his own confusion. He would find no clearing there unless he discovered its prerequisites inside himself. He would have to create his own clearing – even in the jungle of the city.

There used to be three realities: human beings, nature and God. First God disappeared – the great Other who guaranteed sufficient space, the one who opened the greatest room to move. That left nature as the truth of the Other. But if nature too becomes imbued with human enterprise, people end up encountering no more than the unmysterious product of their own activity. Then there remains the worldwide web of civilization, with its people who are ever more alike. In the past, the forests symbolized the mystery, fatefulness and immensity of a world that was still quite different from oneself. Those forests have now become the jungle of human affairs and symbols – which is why we are approaching the age of global claustrophobia.

In this global jungle we again need a kind of clearing. Much as openings used to be made in the forest of first nature, we must now – as Thoreau learned – create clearings in the 'second', man-made nature.

9
False Glows

What does such a clearing in the jungle mean?

Of course, I am now looking at things from the point of view of the individual. The point here is not to discuss which policies (the opening of markets to Third World products, regulation of financial markets, minimum social and ecological standards for world trade, environmental taxation, and so on) might rationally be pursued by states, organizations or citizens' groups against disastrous aspects of globalization, or which political strategies and forms of resistance might be required for that end. The issue, rather, is the life of individuals in a context of global networking, which, as we know, penetrates them so deeply that they feel powerless and insignificant. Naturally, one individual can make little impact externally, but by acting upon oneself it is possible to screen off negative external effects. It is not only the structure of the totality that determines whether life is liveable here and now; the individual also has a certain scope for action. That does not necessarily mean that he or she will use it, or still less enlarge it with ingenuity and self-confidence, but that is the only way to make a clearing of one's own in the jungle of the social.

If we understand the clearing as the space which the individual needs to be an individual, and if we then reflect on how self-willed individuality can be preserved in the face of society, we soon find ourselves inside the story of an old depression.

For more than 200 years people have been complaining that society stifles individual existence; talk of the crisis or even tragedy of culture dates back that long. In this light, the excessive demands of globalization represent the further development of a power which, at least since Rousseau, has been known as social alienation or estrangement. The social totality that caused pain and suffering to the likes of Rousseau, and about which they constantly complained, was still locally circumscribed. Modern globalization has expanded that totality and removed the limits around it. Yet there is still a feeling of crisis about the relationship of the individual to society, and so it makes sense to recall what people used to complain about, which solutions to the crisis they devised, which clearings they sought, and which false lights they have already pursued.

Let us begin with Rousseau, the founder of the modern tradition of cultural criticism within which the new discontent with globalization also stands.

One summer's day in 1749, Rousseau set off from Paris to visit his friend Diderot in prison. On the way he stopped to rest beneath a tree and experienced his great 'inspiration':

> Oh Sir, if I had ever been able to write a quarter of what I saw and felt under that tree, how clearly I would have made all the contradictions of the social system seen, with what strength I would have exposed all the abuses of our institutions, with what simplicity I would have demonstrated that man is naturally good and that it is from these institutions alone that men become wicked.[1]

At that moment Rousseau was overwhelmed by an intoxicating certainty: the individual and all his potential riches are the truth; society outside is governed by a mechanism of deception; society robs the individual of his truth and his liveliness, alienating him from himself. On that summer's day in 1749, beneath the shade of a tree, the theory of alienation that still casts its shadows over contemporary discourse came into being. Since Rousseau we have been convinced that a certain kind of socialization mutilates individual human beings and entangles them in falsehood. This philosophically well-versed discontent with society looks back and discovers a long prehistory. For Rousseau it begins with the fact that people separate themselves off from one another in the pursuit of property and possessions. Property relations are the cause of competition, hierarchies, animosities, mutual distrust, disguises and deception – in short, the whole culture of a society that Rousseau rejects. It is easy to criticize, but how can one find – at least for oneself – a way out of this wretched condition?

Rousseau would not have had such resonance down to the present day if he had sent us on distant journeys: into obscure pasts, exotic lands or an unpredictable future. No, Rousseau pointed to the ostensibly shortest of all journeys: go inside yourselves, he said, and there you will find everything. We are supposed to throw off burdens of knowledge and habits that involve us in the system. It is as if we had only to pull ourselves together to tear the social web in which we were entangled. Rousseau's thought preserves the tempting suggestion that we can escape our wretched condition through the joy of being an individual. The door through which we can reach home is still open.

Jean-Jacques Rousseau redrew the configuration of inner and outer. The inner was hugely enhanced: it was where we had to assume that real life was to be found, while the social exterior appeared by contrast as a soulless mechanism. With consequences for our culture that could not have been foreseen, Rousseau polemically turned the 'truth' of the inner against the outer. In this opposition, the outer inevitably appeared as the world of estrangement. Later philosophies of alienation put us off with grand social projects of emancipation, or else postulated the total blindness of an administered society. But for Rousseau there was what Adorno denied: the possibility of a real life in the midst of falsehood. Individual lives can succeed even if the social totality stands in their way. All I need to do is come back to myself, here and now. But where do I end up when I come to myself?

Rousseau spoke of his 'self' as of a buried treasure brimming with hopes and promises of happiness, and a little later Goethe had his Werther cry out in similar vein: 'I withdraw into myself and discover a world.'[2] Here we see the elation of a release from role pressure, prejudice and pent-up tension. Suddenly it becomes possible again to be on friendly terms with yourself, even with your own body: you want to embrace

nature; imaginative powers become active and bring everything to life; courage to face the present awakens with the
surge of creative freedom; you feel driven to act, and strong
enough to withstand the pressure of history. Freedom is the
ability to begin again; it is the protean nature which prevents
total absorption into social roles. The free self to which
Rousseau wished to return is a kind of inner transcendence, a
prerequisite for creative action and hence for any incarnation
of the sense of possibility.

Rousseau's 'discovery' that there can be a clearing within
ourselves, is not obsolete: it still encourages us to uncover our
own potential.

But let us not forget the drama that can result – and in
Rousseau's case did result – from the experience of our own
freedom. Once we become familiar with freedom and its
creative unpredictability, we can no longer feel safe at the
thought that out there the freedom of many others constitutes
a huge field of unpredictability. Joy in our own freedom
becomes fear of the many freedoms of others; Rousseau grew
panic-stricken about it. He therefore dreamed of society as
a great Communion with one heart and one soul, where
the freedoms of the many would be absorbed into one
great freedom in which everyone was in harmony. One great
freedom – what is that supposed to be? It can only be
the freedom I know from within, and I know only one such
freedom: mine.

This longing for society as a great Communion ends in the
revocation of freedom: society must cease to be something
outer; it must become a single realm of the inner. The danger
lurking in such fantasies has been known to us since
Robespierre practised his terrorism of virtue in the name of
Rousseau's *volonté générale*. Politicization of the demand for
unity, which no longer tolerates any externality (that is, any
difference), inevitably leads to the prison of totalitarianism.

Discovery of ourselves and our freedom is therefore not the
end of the story; we also have to endure the freedom of the
many and the unpredictability that comes with it. Those who
are incapable of this will continue to revel in their own
freedom and, like Rousseau, dream of a great Communion
where there are no longer many freedoms but only the sound
of hearts beating in unison. In comparison with the burning

sense of one's own identity, society appears to be a space of cold difference. But those who want freedom together with the freedom of others have to accept difference, to give up the dream of hearts beating in unison.

For a long time Rousseau's schema saw living subjective spirit pitted against a social mechanism which, though productive in its totality, reduced individuals to the level of mere cogs. But we also find in Rousseau the classical formulation of a way out, where the Archimedean point is the 'true' self beneath the mask of the social subject. Everyone can discover this clearing within themselves, and it is the point from which the social totality itself can be revolutionized. The true revolution is the revolution of the soul. The dynamic centre lies within.

The second classical model for the overcoming of crises, the one associated with Marx, points in an opposite direction.

For Marx the momentum for change lies outside, in the social process. Whereas Rousseau wanted to mobilize the true self against the social mechanism, Marx wants to use that mechanism for the purpose of liberation. The social process, however, cannot be dealt with by means of moral and aesthetic demands; it is necessary to discover and politically instrumentalize its immanent contradictions. It is well known that for Marx the fundamental contradiction was between capital and labour, product and producer. The critique of social alienation was the Rousseauan legacy in this theory. But, since Rousseau, an industrial plant had pushed itself between the individual and the social whole. And so, Marx came up with the idea that history worked in the manner of a machine: 'What the bourgeoisie produces, above all, are its own grave-diggers. Its fall and the victory of the proletariat are equally inevitable.'[3]

Marx discovers outside – in history, in the social process – the inevitability of liberation; the clearing is not the inner experience of self but the light at the end of the historical tunnel. It is there that humanity will wake up as from a bad dream. But a long time will be needed for that to come to pass, and during that time strategic skill must be employed to link up with the dynamic of history.

The light at the end of the historical tunnel turned out to be a false glow. 'Actually existing socialism' was not the great liberation but a grim and cruel prison, where peoples were terrorized and dictated to by an ideological elite. If 'clearing' means the conquest of freedom and sovereignty, then the Communist road led away from such a clearing. Men and women who believed in that road actually strayed into the undergrowth of a history which, they thought, could be made to conform to their own plans. Faith in the supposedly objective dynamic of history-as-progress was bitterly disappointed. So much for trust in a logic of the outer realm.

As to Rousseau's idea that the true self should become the model for socialization, it has proved to end in hostility towards the freedom of the many. The clearing for one becomes a darkening for others. So much for trust in a logic of the inner realm.

Nevertheless, it is essential that a clearing be found, one which is neither completely inward, as in Rousseau, nor completely outward, as in Marx.

10
Creating Space

I imagine the following. We have come from somewhere and want to go somewhere, but we cannot deny that the conditions in which we find ourselves are dark and confused, as in a forest. We have lost our way – a feeling that is nowadays our normal state. We begin to look for a clearing.

Now, there is the possibility of looking for origins, for the true self, for a point where the false trail began. The danger is that we will go astray, backwards or inwards. Then there is the possibility of walking straight ahead to reach where we think we belong: growth, progress. The danger is that we will go astray, forwards or outwards. The third possibility is to stop at the point of present confusion and, without worrying about origin or destination, to cut a clearing there. Clearing: a way of living that is provisionally possible; residence in the midst of confusion; the triumph of being able to begin here and now; an open place with a view of the overarching sky, surrounded by the forest of civilization that is nevertheless kept at a distance.

The calm keeping of distance assumes that we realize that history as a whole is not heading for a sinister destination. History is not a journey for which we can miss the connection, as on a railway. History has already arrived; at every moment it is already finished. As to long-term plans, we must always be prepared for them to turn out differently from how we anticipated. In the jungle of history no plan is ever achieved without deflection. History is the unintended result of count-less individual intentions, which deflect one another as they intersect and intertwine. There is only ever action at close quarters, with restricted view, a mixture of accidents and com-promises, madness, cleverness and force of habit. Instead of making history, man is entangled in various histories and reacts to them in such a way that new ones are continually emerging. History is a teeming mass of histories, and is there-fore notoriously obscure.

Here, to cut a clearing is to discover and energetically hold on to a history of one's own amid the teeming mass of histo-ries; to keep weaving one's own history, in the knowledge that it is entangled in the confusion of many histories and will in the end be lost. This means bidding farewell to the illusion that there is a clearing that might serve as a commanding height, from which history as a whole might be kept on course.

To cut a clearing means to keep cultivating modes of action and thought that refuse simple adaptation to the globalist hysteria: modes such as deceleration, obstinacy, sense of direc-tion, ability to switch off, inaccessibility. Let us not forget, however, that there is no such thing as non-communication in the network of communication, for it too is a communicative act. Everyone who owns a mobile phone knows that: you are always there, and it is up to you to explain why you could not be reached. But you can switch it off, you can be the one who decides whether or not to put it on standby. Permanent availability – the dream of the communications industry – is thought of as progress, but in the past it was only 'domestics' who had to be permanently on call; maybe the idea is for the communications network to employ us as its servants. Openness on all sides and permanent readiness to communi-cate are what is suggested to us. But this overlooks the fact that our minds as well as our bodies need immune defences; we cannot let everything in, only as much as we can usefully convert. The logic of global networking tends to break down cultural immunity, removing any effective system to filter the flood of information. Filtering is possible only if we know what we want and what we need. To avoid bending to the communication pressure, we would have to give up the ambition to keep abreast of the times and at the head of any movement. It is already almost a privilege not to be on the Web, just as it is to enjoy near vision rather than tele-vision. 'We must again become good neighbours of the closest things around us', said Nietzsche.[1]

But has this unburdening, this contemplation of the closest things, not already become a fashion? Is the human type who finds such manoeuvres difficult, who becomes burdened with knowledge and responsibility, not thought to be regressing? Has our civilization not been turning into a society of (increasingly 'single') end-users who, standing outside the chain of generations, only look after themselves? Who still bothers about a sense of social responsibility, instead of just consuming what society has to offer? Is this not the age of keeping things simple and travelling light? One fashionable injunction – to 'clear out our lives' – presupposes that the junk piling up at home has a spiritual equivalent in the ballast of questionable ideas and knowledge which prevent us from happily taking control of our lives. And, indeed, advice to get rid of the dross conceals an attack on the old European tradition of serious consideration and willingness to assume the burden of complexity. Such conceptions, which recall from afar Nietzsche's blow for freedom against historicism, will in their way indeed hack out a clearing in the modern jungle of overtaxing demands.

Initially it looks as if their main concern is with development of the individual. But it is an individual constructed according to the model of the self-confident consumer, who knows how to choose from all the goods on offer and is careful not to become too dependent on people or things. Life's guiding principles here are disposability and non-returnability, fast consumption and fast throughput. These common terms, all derived from the sphere of the market, promote a gutted individuality and are therefore part of the problem that they are supposed to solve. For the significance of the individual is not reducible to the function of end-user. What does it mean to become an individual?

Shortly before his death, Wilhelm von Humboldt wrote in a letter: 'Someone who can say at the end of his life "I have fathomed as much of the world as I could and have converted it into my humanity" has achieved his goal.'[2] To convert the world into the 'humanity' one is – that is precisely the point.

The strength for such conversion should be the measure of one's curiosity about the world. But, as we know, that is easier said than done, when all-round openness and constant willingness to communicate are held up as the norm. The prerequisite for individual development, in Humboldt's sense of the term, is that one has an idea of what one's life should look like, and that this idea helps one to sort through and process the wealth of stimuli and information coming from outside. This strength to shape one's own life used to be known as the fruit of 'education', by contrast to mere 'training'. Education is the development of the individual as an end in itself; training is a means of preparing the individual for the labour process. Of course, we need both the one and the other. Whereas training links us into an external network, education is the development of the network that each individual is for himself or herself.

It was noted long ago that the modern world requires training but does not encourage education. In Schiller we read:

> Everlastingly chained to a single little fragment of the Whole, man himself develops into nothing but a fragment; everlastingly in his ear the monotonous sound of the wheel that he turns, he never develops the harmony of his being, and instead of putting the stamp of humanity upon his own nature, he becomes nothing more than the imprint of his occupation or of his specialized knowledge.[3]

About the turn of the nineteenth century, the idealist vision of education suggested that, despite everything, the individual might become a whole, a totality in miniature. Schiller sought to achieve this through the 'aesthetic education of man'. In the enjoyment or production of art, he wrote, man experiences a self-contained context that both demands and cultivates the human powers – senses, understanding and feeling. It is a non-purposive context and is not in the service of anything else; it carries its meaning and purpose within itself and may therefore be said to possess sovereign dignity. In the miniature totality of the realm of art, man plays with his 'essential powers' and compensates for the mutilation inflicted upon him by the division of labour in society.

Schiller was aware that such aesthetic education cannot have a wide social impact and is not suitable as a political strategy; nor did he expect aesthetics to bring about a fundamental change in the difficult reality of his time. It was enough for him that the aesthetic sense offered some protection against the devastating effects of that reality. The 'aesthetic education of man' was a defensive concept. It aimed to reach the few individuals who, wanting more than to work, consume and function, sought to develop the 'humanity' within themselves – first in the experimental theatre of art, and then perhaps in the rest of life. As a defensive concept for the few, it is still relevant today. From art it is possible to learn how important are limits and boundaries, since the wish for form creates a strictly delimited zone of meaning that we call art and differentiate from the rest of everyday reality. Artworks, if worthy of the name, are formally self-contained and make it possible to experience broad expanses within a narrowly defined space. They show a rich abundance within precise boundaries, and can therefore be a school for a life that does not wish to dissipate its energies. The aesthetic sense gives a foretaste of the joy of a sovereign existence.

Globality brings us into contact with more and more reality, and it is difficult to preserve any sovereignty there. A sovereign individual would be one who decided even his or her entanglements and points of contact with reality. Such sovereignty presupposes an existential power of judgement: that is, a feel for what is really of concern to us; a capacity to distinguish degrees of urgency and the range of our action. Globalization hysteria is due to the fact that this capacity to distinguish between the existentially close and the existentially distant has been impaired or even destroyed. This is what Goethe had in mind when he said that 'it is always a misfortune when a man is induced to strive for something with which he cannot associate himself through some regular spontaneous activity'.[4] So long as one is under the sway of the idea that there is no genuine life amid falsehood, it is very difficult to find the courage to work out which 'spontaneous activity' is right for oneself.

Those who wish to create their clearing in the jungle of the social and the dense undergrowth of global communication will not be able to manage it without drawing existentially astute boundaries. Those who wish to shape a life of their own must know the point at which it ceases to be amenable to formatting and in-forming. When Karl Jaspers discussed the difficulty of maintaining an autonomous will, he compared it to life on 'a narrow mountain ridge', from which it is possible to fall 'on one side into the mere enterprise, or, on the other, into a life devoid of reality side by side with the enterprise'.[5]

Under the sign of global communication, world time forces its way into individual lifetime. It is as if world time were devouring the individual lifetime, as if the latter no longer mattered. But it does matter. It is there that everything is decided for the individual. A geometrical absurdity – namely, that the larger circle fits into the smaller circle of an individual life without shattering it – can become actual in terms of practical life. Johann Peter Hebel tells of this in his wonderful story 'Unexpected Reunion'. A betrothed couple at Falun in Sweden. Early on their wedding day, the bridegroom went once more down the mine and never came back. The bride waited and waited, for fifty years. Hebel continues:

> In the meantime, the city of Lisbon in Portugal was destroyed by an earthquake, the Seven Years War came and went, the Emperor Francis I died, the Jesuits were dissolved. Poland was partitioned, the Empress Maria Theresa died, and Struensee was executed, and America became independent, and the combined French and Spanish force failed to take Gibraltar. . . . The French Revolution came and the long war began, and the Emperor Leopold II too was buried. Napoleon defeated Prussia, the English bombarded Copenhagen, and the farmers sowed and reaped.[6]

In the meantime the bride waited, until in the end, fifty years later, some miners found the corpse of the groom, untouched by decay, in the vitriol water of a collapsed underground passage. The span of a life ended with this unexpected reunion.

A life story, we say, fits into world history; it is contained within it. But here the opposite happens: world history is inserted into a life story, between parentheses, loosely associated through that inimitable 'in the meantime'. I wish that we could keep globalization similarly at a distance, between parentheses, only loosely associated with our own lives through an 'in the meantime'.

But perhaps such a way of looking at the things of life demands the cleverness of those giants who created the first clearings in the forest.

Notes

Chapter 1 First Nature, Second Nature

1 Arthur Schopenhauer, *The World as Will and Representation*, trans. E. F. J. Payne, vol. 2, *'Supplements to the First Book'*, New York: Dover Publications, 1958, p. 3.
2 K. Marx and F. Engels, 'Manifesto of the Communist Party', in Karl Marx, *The Revolutions of 1848*, London: Penguin Books/New Left Review, 1973, p. 70; Adalbert Stifter, *Der Nachsommer*, Düsseldorf: L. Schwann, 1949, pp. 738–9.

Chapter 3 Globalism

1 Marx and Engels, 'Manifesto of the Communist Party', p. 70.
2 Adam Smith, *The Theory of Moral Sentiments*, ed. D. D. Raphael and A. L. Macfie, Oxford: Clarendon Press, 1976, p. 231.
3 See Joseph Stiglitz, *Globalization and its Discontents*, London: Allen Lane, 2002.

Chapter 4 Making Enemies

1 Friedrich Nietzsche, *Human, All Too Human*, trans. Marion Faber, with Stephen Lehmann, Lincoln, NE: University of Nebraska Press, 1984, pp. 30–1.

2 F. Hegel, *The Logic of Hegel*, trans. William Wallace, 2nd edn, London: Oxford University Press, 1892, ch. 92, p. 173. [German terms added by present translator.]
3 From F. Schiller's original *Ode to Joy* (1785), later adapted for Beethoven's Ninth Symphony.
4 Dante Alighieri, *The Divine Comedy*, trans. C. H. Sisson, London: Pan Books, 1981, p. 449.

Chapter 5 World Peace?

1 'Perpetual Peace: A Philosophical Sketch', in *Kant: Political Writings*, trans. H. B. Nisbet, ed. Hans Reiss, Cambridge: Cambridge University Press, 1991, pp. 93–130.
2 Ibid., p. 106.
3 Ibid., p. 103.
4 Ibid., p. 114.
5 Ibid., pp. 129–30.
6 Ibid., p. 100.
7 Ibid., p. 114.
8 Ibid., p. 112.

Chapter 6 The Global and the Other Totality

1 Friedrich Schiller, 'The Stage as a Moral Institution' (1784), in *Essays Aesthetical and Philosophical*, London: George Bell & Sons, 1910, p. 334.

Chapter 7 The Individual and the Immune System

1 Johann Wolfgang von Goethe, *Wilhelm Meister's Years of Apprenticeship*, trans. H. M. Waidson, vol. 2, books 4–6, London: John Calder, 1978, p. 174. [Parentheses added.]
2 Walter Benjamin, 'The Work of Art in the Age of Mechanical Reproduction', in *Illuminations*, trans. Harry Zohn, New York: Schocken Books, 1968, p. 222.
3 Heinrich Heine, *Gesamtausgabe*, vol. 14/1, *Lutezia II. Berichte über Politik, Kunst und Volksleben*, Hamburg: Hoffmann und Campe, 1990, p. 58.

Chapter 8   Jungle and Clearing

1   Giambattista Vico, *The New Science*, trans. T. G. Bergin and M. H. Fisch, Ithaca, NY: Cornell University Press, 1968, p. 14.
2   *The Epic of Gilgamesh*, trans. N. K. Sandars, London: Penguin, 1972, p. 72.
3   Ibid.
4   Dante, *The Divine Comedy*, p. 47.
5   R. Descartes, 'Discourse on the Method of Rightly Conducting the Reason', in *The Philosophical Works of Descartes*, trans. Elizabeth S. Haldane and G. R. T. Ross, London: Cambridge University Press, 1931, p. 96.
6   Ibid., p. 88.
7   Henry David Thoreau, *Walden*, Boston, MA: Houghton Mifflin, 2000, pp. 101, 106.

Chapter 9   False Glows

1   J.-J. Rousseau, 'Letter to M. de Malesherbes' [12 January 1762], in *The Collected Writings of Rousseau*, vol. 5, trans. Christopher Kelly, Hanover, NH: University Press of New England, 1995, p. 575.
2   Johann Wolfgang von Goethe, *The Sorrows of Young Werther*, trans. Michael Hulse, London: Penguin, 1989, p. 31.
3   Marx and Engels, 'Manifesto of the Communist Party', p. 79.

Chapter 10   Creating Space

1   Friedrich Nietzsche, *Werke*, vol. 4iv, ed. G. Colli and M. Montinari, Berlin: Walter de Gruyter, 1969, p. 356.
2   Wilhelm von Humboldt, *Gesammelte Schriften*, Berlin: Walter de Gruyter, 1968. [English translation provided is translator's own.]
3   Friedrich Schiller, *On the Aesthetic Education of Man*, trans. Elizabeth M. Wilkinson and L. A. Willoughby, Oxford: Clarendon Press, 1967, p. 35.
4   Goethe, *Wilhelm Meister's Years of Apprenticeship*, p. 174.
5   Karl Jaspers, *Man in the Modern Age*, trans. Eden and Cedar Paul, London: Routledge & Kegan Paul, 1951, p. 198.
6   Johann Peter Hebel, 'Unexpected Reunion', in *The Treasure Chest*, trans. John Hibberd, London: Penguin, 1994, p. 26.

Index

CPSIA information can be obtained
at www.ICGtesting.com
Printed in the USA
BVHW041642021120
592333BV00009B/124